Springer Tracts in Natural Philosophy

Volume 22

William Alan Day

The Thermodynamics of Simple Materials with Fading Memory

Springer-Verlag New York Heidelberg Berlin 1972

William Alan Day
The University of Oxford

AMS Subject Classifications (1970)
Primary 73 A 05, 73 B 30; Secondary 80 A 05, 80 A 10

ISBN 0-387-05704-8 Springer-Verlag New York Heidelberg Berlin
ISBN 3-540-05704-8 Springer-Verlag Berlin Heidelberg New York

For Enid

Preface

This Tract gives an account of certain recent attempts to construct a satisfactory theory of thermodynamics for materials which have a memory for the past. Naturally it draws heavily on the writings of those who have made significant contributions to the field.

I am particularly grateful to Professor C. A. Truesdell of The Johns Hopkins University for his invitation to write the Tract and to Professor A. E. Green of Oxford for his comments on various parts of the manuscript.

Hertford College, Oxford
December 1971 W. A. Day

Contents

Introduction

In the decade ending in 1971 much research has been undertaken in an attempt to produce an acceptable theory of the macroscopic thermo-mechanical behaviour of materials which have a memory for the past, or at least some aspects of the past. The theory sought is required to be acceptable in the sense that it applies to an extensive class of materials with non-linear response and proceeds by making rigorous mathematical deductions from clearly defined assumptions.

Any theory of thermodynamics must resolve two separate matters. To begin with it must face the question as to what is a physically sound, sufficiently general and mathematically precise statement of the irreversible nature of the processes involved; in other words it must decide on the appropriate form of the second law. Once an answer to this question has been hazarded there remains the technical matter of elaborating the consequences of the second law. Bearing in mind the paucity of our knowledge in genuine thermodynamics, as opposed to thermostatics or pure mechanics, and bearing in mind too the ambitious scope of the applications we have in mind, it is perhaps unreasonable to claim that the first question has been answered fully and sufficiently generally for all time[1]. If, instead, we take the view that there are reasonable grounds for doubt about the formulation of the second law then it seems prudent to proceed by examining the various forms which have been proposed for the second law, elaborating the consequences of each of the proposed forms and by attempting to decide whether or not these consequences are physically reasonable.

The present monograph is devoted to an account of just two approaches to thermodynamics. They start from different forms of the second law but it turns out that for a great many materials of interest they lead to the same or closely similar results. The earlier of the two

[1] Cf., for example, section 6.3 where it is pointed out that the monotonicity of the one-dimensional relaxation functions of linear viscoelastic materials, which is verified experimentally, is not a consequence of either of the two forms of the second law used here.

theories is due to B. D. Coleman who was the first to construct a rational and exact theory of thermodynamics. It was Coleman who first discerned and gave proofs of the salient results in the thermodynamics of materials with memory. Coleman's theory takes the Clausius-Duhem inequality to be the expression of the second law and it provides a powerful systematic method for reducing constitutive equations to forms compatible with thermodynamics. Many applications of the theory have been made subsequently both by Coleman himself and by other writers.

The statement of the Clausius-Duhem inequality involves the entropy of the body, which means that in Coleman's theory one must have the concept of entropy to hand at the very outset. The second theory described here arises out of two papers[1] devoted to deriving similar results to Coleman's from a different starting point. The essential features of this theory are that the starting form of the second law is, in general, less restrictive than the Clausius-Duhem inequality, so that one can usually prove less from it, and that the entropy concept does not enter into the starting formulation. Instead it is shown how for many materials this form of the second law enables one to construct the entropy, which now enters the theory as a derived concept, and which can be shown to have many properties in common with the entropy in Coleman's theory. Against this advantage, however, one must set the disadvantage that the technical problem of finding restrictions on constitutive equations is almost always much less straightforward and systematic than it is in Coleman's theory.

The monograph makes no attempt to cover completely all the research carried out in the last ten years in continuum thermodynamics; it concentrates on a rather narrow part of the field and a great deal has been omitted which should appear in a comprehensive account. In particular I do not discuss in any detail Müller's[2] proposed generalisation of the Clausius-Duhem inequality or Meixner's[3] proposal for the second law. Nor do I consider, for example, the work of R. M. Bowen, A. E. Green, M.E. Gurtin, I. Muller, C. Truesdell and others on mixtures, nor the careful axiomatic treatment of the first two laws of thermodynamics by M.E. Gurtin and W.O. Williams nor the work of A.E. Green and P.M. Naghdi on plasticity nor that of D.R. Owen on the thermodynamics of rate-independent materials nor that of A.E. Green and N. Laws on the use of global, rather than local, forms of the second law.

It is possible to present both the theories of this monograph as formal mathematical structures starting from a small number of consistent and

[1] See [29, 30].
[2] See [60].
[3] See [58].

precisely stated axioms and proceeding by way of rigorous proofs and constructions and so on and it seems worth emphasising that this is so, for it is not so, or at least has not been shown to be so, for some approaches to thermodynamics. Nevertheless I have elected not to write in quite such a formal vein and I have tried instead to emphasise the important physical assumptions which are made rather than the mathematical technicalities. There are places in the text, however, where it is vital not to be too cavalier about assumptions of smoothness and the like and these points are discussed in the detail they deserve.

The plan of the monograph is as follows. Chapter 1 is devoted to some preliminaries; it begins by reviewing certain results of vector and tensor analysis and certain results about the line integrals of vector fields. It then outlines those facts about kinematics and the balance laws for mass, momentum, moment of momentum and energy which are relevant to our purpose and it concludes by introducing the class of simple materials with memory to be studied in the following Chapters. Chapters 2, 3 and 4 are concerned with the more recent theory in which the entropy enters as a derived concept. In Chapter 2 the fundamental thermodynamic inequality, which is a formulation of the second law, is introduced and used to derive results about heat conduction and to recover, in a rather general context, certain classical results on the conversion of heat into mechanical work. Chapter 3 is devoted entirely to constructing the entropy and to determining its properties; it turns out for an extensive class of materials satisfying the thermodynamic inequality that provided the materials have fading memory and are well-behaved under the operation of retarding processes then the entropy can be constructed not only in equilibrium but away from equilibrium as well. Chapter 4 illustrates the arguments and results of Chapter 3 by constructing the entropy explicitly for certain materials.

Coleman's theory is presented, with some slight differences in detail from Coleman's original treatment, in Chapter 5. The most important part of the Chapter is to be found in sections 5.3 and 5.4 which treat the thermodynamics of simple materials with fading memory and instantaneous elastic response. These sections reproduce most of the results of the fundamental memoirs [8, 9]. The Chapter concludes by discussing the connection with the theory of Chapters 2, 3 and 4.

One of the results to emerge from Chapter 5 is that the work done around any closed isothermal strain path starting from equilibrium cannot be negative. This result necessarily imposes restrictions on the material and in particular it restricts the behaviour of the relaxation function of a linear viscoelastic material; these restrictions and certain related restrictions are investigated in detail in Chapter 6.

CHAPTER 1

Preliminaries

1.1 Vector and Tensor Analysis

We begin by rapidly surveying the definitions and results we need from vector and tensor analysis. We write $\boldsymbol{v} \cdot \boldsymbol{w}$ for the *scalar product* of the vectors $\boldsymbol{v}, \boldsymbol{w}$ in three-dimensional euclidean space and $|\boldsymbol{v}| = (\boldsymbol{v} \cdot \boldsymbol{v})^{\frac{1}{2}}$ for the *norm* or length of $\boldsymbol{v}$. The term *tensor* (second order) stands for any linear transformation of euclidean space into itself. We write det $\boldsymbol{L}$ for the determinant of the tensor $\boldsymbol{L}$ and trace $\boldsymbol{L}$ for its trace. Its *transpose* is the tensor $\boldsymbol{L}^T$ defined by the condition $\boldsymbol{w} \cdot \boldsymbol{L}\boldsymbol{v} = \boldsymbol{v} \cdot \boldsymbol{L}^T \boldsymbol{w}$, holding for every pair of vectors $\boldsymbol{v}, \boldsymbol{w}$. The tensor $\boldsymbol{L}$ is said to be *symmetric* if $\boldsymbol{L}^T = \boldsymbol{L}$, *skew-symmetric* if $\boldsymbol{L}^T = -\boldsymbol{L}$ and *orthogonal* if it preserves scalar products, that is if $(\boldsymbol{L}\boldsymbol{v}) \cdot (\boldsymbol{L}\boldsymbol{w}) = \boldsymbol{v} \cdot \boldsymbol{w}$. The orthogonality of $\boldsymbol{L}$ is equivalent to the conditions $\boldsymbol{L}\boldsymbol{L}^T = \boldsymbol{L}^T \boldsymbol{L} = \boldsymbol{I}$, where $\boldsymbol{I}$ is the identity tensor. The tensor $\boldsymbol{L}$ is *positive semi-definite* if $\boldsymbol{v} \cdot \boldsymbol{L}\boldsymbol{v} \geqslant 0$ for every vector $\boldsymbol{v}$ and *positive definite* if the stronger condition $\boldsymbol{v} \cdot \boldsymbol{L}\boldsymbol{v} > 0$ holds whenever $\boldsymbol{v} \neq \boldsymbol{0}$. If the tensor $\boldsymbol{F}$ has a positive determinant it admits the *polar decomposition*

$$\boldsymbol{F} = \boldsymbol{R}\boldsymbol{U} \tag{1.1.1}$$

into the product of an orthogonal tensor $\boldsymbol{R}$ with a symmetric and positive definite tensor $\boldsymbol{U}$. There is only one choice of $\boldsymbol{R}$ and $\boldsymbol{U}$ meeting these conditions. The *tensor product* of the vectors $\boldsymbol{v}, \boldsymbol{w}$ is the tensor $\boldsymbol{v} \otimes \boldsymbol{w}$ defined by

$$(\boldsymbol{v} \otimes \boldsymbol{w})\,\boldsymbol{u} = (\boldsymbol{u} \cdot \boldsymbol{w})\,\boldsymbol{v}\,. \tag{1.1.2}$$

The collection of all (second order) tensors can itself be regarded as a nine-dimensional vector space having the natural *scalar product*

$$\boldsymbol{L} \cdot \boldsymbol{M} = \operatorname{trace} \boldsymbol{L}\boldsymbol{M}^T \tag{1.1.3}$$

and the *norm*

$$|\boldsymbol{L}| = (\boldsymbol{L} \cdot \boldsymbol{L})^{\frac{1}{2}}\,. \tag{1.1.4}$$

With this definition of the scalar product the *triangle inequality*

$$|\boldsymbol{L}+\boldsymbol{M}| \leqslant |\boldsymbol{L}|+|\boldsymbol{M}| \tag{1.1.5}$$

and the *Schwarz inequality*

$$|\boldsymbol{L}\cdot\boldsymbol{M}| \leqslant |\boldsymbol{L}|\,|\boldsymbol{M}| \tag{1.1.6}$$

hold. The symmetric tensors can be regarded as belonging to a six-dimensional subspace of the nine-dimensional space. Later on we shall need to consider linear transformations of the nine-dimensional space into itself; we shall call them *fourth order tensors.*

Its is of course possible to introduce a rectangular cartesian coordinate system defined by three mutually orthogonal unit vectors $\boldsymbol{e}_1, \boldsymbol{e}_2, \boldsymbol{e}_3$, and occasionally it is convenient to do so for then a vector $\boldsymbol{v}$, a second order tensor $\boldsymbol{L}$ and a fourth order tensor $\mathscr{L}$ have concrete representations in terms of their scalar components relative to the coordinate system:

$$v_i = \boldsymbol{v}\cdot\boldsymbol{e}_i \qquad (i=1,2,3), \tag{1.1.7}$$

$$L_{ij} = \boldsymbol{e}_i\cdot(\boldsymbol{L}\boldsymbol{e}_j) \qquad (i,j=1,2,3), \tag{1.1.8}$$

$$\mathscr{L}_{ijkl} = (\boldsymbol{e}_i\otimes\boldsymbol{e}_j)\cdot\{\mathscr{L}(\boldsymbol{e}_k\otimes\boldsymbol{e}_l)\} \qquad (i,j,k,l=1,2,3) \tag{1.1.9}$$

where the scalar product occurring in (1.1.9) is the one defined in (1.1.3). All the operations we perform have straightforward interpretations in terms of cartesian components.

We shall need to differentiate scalar, vector and tensor fields defined on an open subset $\mathscr{U}$ of euclidean space. The scalar field $\varphi(\cdot)$ on $\mathscr{U}$ is said to be differentiable if there is a vector field $\boldsymbol{w}(\cdot)$ on $\mathscr{U}$ such that

$$\lim_{\boldsymbol{x}\to\boldsymbol{x}_0} \frac{|\varphi(\boldsymbol{x})-\varphi(\boldsymbol{x}_0)-\boldsymbol{w}(\boldsymbol{x}_0)\cdot(\boldsymbol{x}-\boldsymbol{x}_0)|}{|\boldsymbol{x}-\boldsymbol{x}_0|} = 0 \tag{1.1.10}$$

for every point $\boldsymbol{x}_0$ in $\mathscr{U}$. If this is so the vector field $\boldsymbol{w}(\cdot)$ is unique; we call it the *gradient* of $\varphi(\cdot)$ and write

$$\boldsymbol{w}(\cdot) = \operatorname{grad}\varphi(\cdot)\,. \tag{1.1.11}$$

The gradient of a vector field $\boldsymbol{v}(\cdot)$ is the tensor field $\operatorname{grad}\boldsymbol{v}(\cdot)$ defined by the condition

$$\{\operatorname{grad}\boldsymbol{v}(\cdot)\}^T\boldsymbol{c} = \operatorname{grad}(\boldsymbol{c}\cdot\boldsymbol{v}(\cdot)) \tag{1.1.12}$$

for every constant vector $\boldsymbol{c}$ and the *divergence* of $\boldsymbol{v}(\cdot)$ is the scalar field

$$\operatorname{div}\boldsymbol{v}(\cdot) = \operatorname{trace}(\operatorname{grad}\boldsymbol{v}(\cdot))\,. \tag{1.1.13}$$

The *divergence* of the tensor field $\boldsymbol{L}(\cdot)$ is the vector field $\operatorname{div}\boldsymbol{L}(\cdot)$ for which

$$\boldsymbol{c}\cdot\operatorname{div}\boldsymbol{L}(\cdot) = \operatorname{div}(\boldsymbol{L}^T(\cdot)\boldsymbol{c}) \tag{1.1.14}$$

for every constant vector $\boldsymbol{c}$. Of course if the scalar field $\varphi(\cdot)$ is twice continuously differentiable then its second gradient

$$\operatorname{grad}^2 \varphi(\cdot) = \operatorname{grad}(\operatorname{grad} \varphi(\cdot)) \tag{1.1.15}$$

is a symmetric tensor field.

If $\mathscr{U}$ happens to be a bounded region whose boundary $\partial\mathscr{U}$ is sufficiently well-behaved the divergence theorem asserts that

$$\int_{\mathscr{U}} \operatorname{div} \boldsymbol{v}(\boldsymbol{x})\, dV(\boldsymbol{x}) = \int_{\partial\mathscr{U}} \boldsymbol{v}(\boldsymbol{x}) \cdot \boldsymbol{n}(\boldsymbol{x})\, d\mathscr{A}(\boldsymbol{x}) \tag{1.1.16}$$

and

$$\int_{\mathscr{U}} \operatorname{div} \boldsymbol{L}(\boldsymbol{x})\, dV(\boldsymbol{x}) = \int_{\partial\mathscr{U}} \boldsymbol{L}(\boldsymbol{x})\, \boldsymbol{n}(\boldsymbol{x})\, d\mathscr{A}(\boldsymbol{x}), \tag{1.1.17}$$

where $\boldsymbol{n}(\cdot)$ is the unit outward normal to $\partial\mathscr{U}$, where $\int_{\mathscr{U}} \dots dV(\boldsymbol{x})$ stands for a volume integration taken over $\mathscr{U}$ and $\int_{\partial\mathscr{U}} \dots d\mathscr{A}(\boldsymbol{x})$ for a surface integration over $\partial\mathscr{U}$.

For the sake of definiteness we have tacitly discussed differentiation for scalar, vector and tensor fields defined only on subsets of three-dimensional euclidean space but all the concepts introduced here carry over with at most trivial modifications to spaces of dimension other than three and later on we shall make use of this fact.

1.2 Paths and Line Integrals

In the preceding section we looked briefly at differentiation for functions defined on an open set $\mathscr{U}$. One way of investigating the properties of vector fields on $\mathscr{U}$ is to look at the line integrals of the fields along various paths in $\mathscr{U}$. For us the term *path* means a function $\boldsymbol{f}(\cdot)$ which associates with each number t a point $\boldsymbol{f}(t)$ in $\mathscr{U}$ and meeting certain requirements which will be stated below. In practice we shall always think of t as "time" and interpret $\boldsymbol{f}(t)$ as the value of some physical quantity at the time t. If $\boldsymbol{f}(\cdot)$ is to be a path it must be continuous and it must start and finish at rest in the sense that there are finite times τ^- and τ^+, say, such that $\boldsymbol{f}(\cdot)$ is constant for all times $t \leqslant \tau^-$ and $\boldsymbol{f}(\cdot)$ is constant for all times $t \geqslant \tau^+$. The constant value $\boldsymbol{f}(-\infty)$ assumed by $\boldsymbol{f}(\cdot)$ on $(-\infty, \tau^-]$ is called the *initial value* of $\boldsymbol{f}(\cdot)$ and the constant value $\boldsymbol{f}(+\infty)$ assumed on $[\tau^+, +\infty)$ is its *final value*. The values $\boldsymbol{f}(-\infty)$ and $\boldsymbol{f}(+\infty)$ do not necessarily coincide but if they do $\boldsymbol{f}(\cdot)$ is called a *closed path*. For technical mathematical reasons we assume too that on the time interval $\tau^- \leqslant t \leqslant \tau^+$ the function $\boldsymbol{f}(\cdot)$ is piecewise C^∞ which just means that the interval can be partitioned by a finite number of times $\tau^- = t_0 < t_1 < \dots < t_{n-1} < t_n = \tau^+$ in such a way that $\boldsymbol{f}(\cdot)$ has derivatives of all orders on each sub-interval $t_m \leqslant t \leqslant t_{m+1}$ $(m=0,\ldots,n-1)$. A schematic representation of a typical path is shown in Fig. 1.

From here on it will be assumed that the set $\mathscr{U}$ is not merely open but is also *connected* in the sense that any two points $\boldsymbol{x}$ and $\boldsymbol{y}$ of $\mathscr{U}$ can be

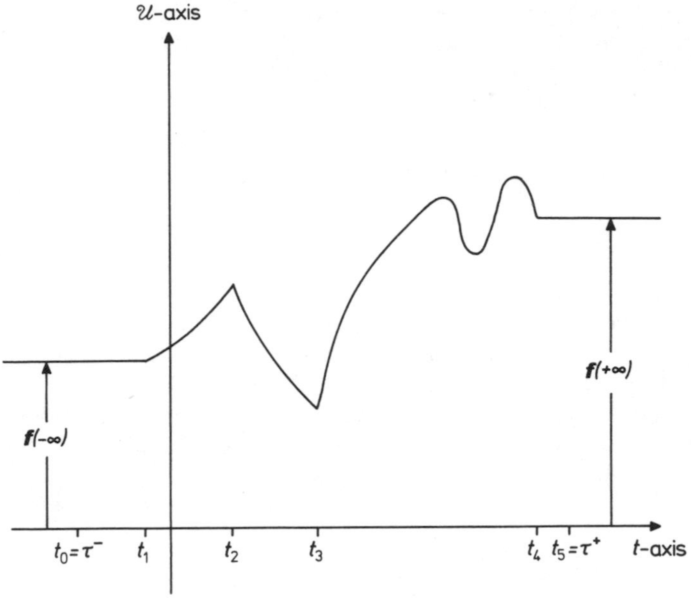

Fig. 1: A path $\boldsymbol{f}(\cdot)$

joined by a path $\boldsymbol{f}(\cdot)$ with initial value $\boldsymbol{f}(-\infty)=\boldsymbol{x}$, with final value $\boldsymbol{f}(+\infty)=\boldsymbol{y}$ and with $\boldsymbol{f}(t)$ lying in $\mathscr{U}$ for every time t. If $\boldsymbol{v}(\cdot)$ is a continuous vector field on $\mathscr{U}$ the *line integral* of $\boldsymbol{v}(\cdot)$ along the path $\boldsymbol{f}(\cdot)$ is

$$I(\boldsymbol{v}(\cdot),\boldsymbol{f}(\cdot))=\int_{-\infty}^{+\infty}\boldsymbol{v}(\boldsymbol{f}(t))\cdot\dot{\boldsymbol{f}}(t)\,dt\,, \tag{1.2.1}$$

where $\dot{\boldsymbol{f}}(\cdot)$ is the derivative of $\boldsymbol{f}(\cdot)$, that is $\dot{\boldsymbol{f}}(t)=(d/dt)\boldsymbol{f}(t)$. Of course the very definition of a path tells us that $\dot{\boldsymbol{f}}(\cdot)$ vanishes outside some finite time interval $\tau^- \leqslant t \leqslant \tau^+$ so that the integration over all values of t in equation (1.2.1) can be replaced by an integration over a finite interval.

In our study of thermodynamics we shall frequently be led to ask if a given vector field $\boldsymbol{v}(\cdot)$ can be expressed as the gradient of some scalar field. It is well known that one way of testing if this is so is by using line integrals: *the vector field* $\boldsymbol{v}(\cdot)$ *can be expressed as the gradient* $\operatorname{grad}\varphi(\cdot)$ *of some scalar field* $\varphi(\cdot)$ *on* $\mathscr{U}$ *if the line integral of* $\boldsymbol{v}(\cdot)$ *around every closed path* $\boldsymbol{f}(\cdot)$ *in* $\mathscr{U}$ *is never negative, that is if* $I(\boldsymbol{v}(\cdot),\boldsymbol{f}(\cdot))\geqslant 0$.

This result can be proved in the following way. To begin with the assumption that the line integral around a closed path is never negative tells us that if $\boldsymbol{g}(\cdot)$ and $\boldsymbol{h}(\cdot)$ are any two paths with the same initial values and the same final values, that is if $\boldsymbol{g}(-\infty)=\boldsymbol{h}(-\infty)$ and $\boldsymbol{g}(+\infty)=\boldsymbol{h}(+\infty)$, then $I(\boldsymbol{v}(\cdot), \boldsymbol{g}(\cdot))=I(\boldsymbol{v}(\cdot), \boldsymbol{h}(\cdot))$. This follows because we can define a closed path $\boldsymbol{f}(\cdot)$ by choosing a time τ^+ in such a way that $\boldsymbol{g}(t)=\boldsymbol{h}(t)$ for all times $t \geqslant \tau^+$ and having done this we can construct $\boldsymbol{f}(\cdot)$ by setting $\boldsymbol{f}(t)=\boldsymbol{g}(t)$ for times $t \leqslant \tau^+$ and $\boldsymbol{f}(t)=\boldsymbol{h}(2\tau^+ - t)$ for times $t \geqslant \tau^+$. The line integral of $\boldsymbol{v}(\cdot)$ along this path is

$$I(\boldsymbol{v}(\cdot), \boldsymbol{f}(\cdot))=I(\boldsymbol{v}(\cdot), \boldsymbol{g}(\cdot))-I(\boldsymbol{v}(\cdot), \boldsymbol{h}(\cdot)).$$

Since $\boldsymbol{f}(\cdot)$ is closed, $I(\boldsymbol{v}(\cdot), \boldsymbol{f}(\cdot)) \geqslant 0$ and consequently

$$I(\boldsymbol{v}(\cdot), \boldsymbol{g}(\cdot)) \geqslant I(\boldsymbol{v}(\cdot), \boldsymbol{h}(\cdot)).$$

If we now interchange the roles of $\boldsymbol{g}(\cdot)$ and $\boldsymbol{h}(\cdot)$ we deduce that $I(\boldsymbol{v}(\cdot), \boldsymbol{g}(\cdot))=I(\boldsymbol{v}(\cdot), \boldsymbol{h}(\cdot))$. This means that a scalar field $\varphi(\cdot)$ can be defined unambiguously by choosing a fixed point $\boldsymbol{x}_0$ in $\mathscr{U}$ and setting

$$\varphi(\boldsymbol{x})=I(\boldsymbol{v}(\cdot), \boldsymbol{g}(\cdot)), \tag{1.2.2}$$

where $\boldsymbol{g}(\cdot)$ is any path with initial value $\boldsymbol{g}(-\infty)=\boldsymbol{x}_0$ and final value $\boldsymbol{g}(+\infty)=\boldsymbol{x}$. It remains to be shown that the gradient of $\varphi(\cdot)$ is equal to $\boldsymbol{v}(\cdot)$ everywhere.

If $\boldsymbol{x}$ is any point of $\mathscr{U}$, the fact that $\mathscr{U}$ is open enables us to choose an open sphere with centre $\boldsymbol{x}$ lying entirely in $\mathscr{U}$. For any point $\boldsymbol{y}$ of this sphere the straight line segment $\boldsymbol{x}+t(\boldsymbol{y}-\boldsymbol{x})$, $0 \leqslant t \leqslant 1$, joining $\boldsymbol{x}$ to $\boldsymbol{y}$ lies in $\mathscr{U}$ and it follows from the definition (1.2.2) of $\varphi(\cdot)$ that

$$\varphi(\boldsymbol{y})-\varphi(\boldsymbol{x})=\int_0^1 \boldsymbol{v}(\boldsymbol{x}+t(\boldsymbol{y}-\boldsymbol{x}))\cdot(\boldsymbol{y}-\boldsymbol{x})\,dt.$$

However, the mean value theorem for integrals tells us that there is some number α in $0 \leqslant \alpha \leqslant 1$, depending on both $\boldsymbol{x}$ and $\boldsymbol{y}$, with the property

$$\varphi(\boldsymbol{y})-\varphi(\boldsymbol{x})=\boldsymbol{v}(\boldsymbol{x}+\alpha(\boldsymbol{y}-\boldsymbol{x}))\cdot(\boldsymbol{y}-\boldsymbol{x})$$

and so

$$\begin{aligned}\frac{|\varphi(\boldsymbol{y})-\varphi(\boldsymbol{x})-\boldsymbol{v}(\boldsymbol{x})\cdot(\boldsymbol{y}-\boldsymbol{x})|}{|\boldsymbol{y}-\boldsymbol{x}|} &= \frac{|(\boldsymbol{v}(\boldsymbol{x}+\alpha(\boldsymbol{y}-\boldsymbol{x}))-\boldsymbol{v}(\boldsymbol{x}))\cdot(\boldsymbol{y}-\boldsymbol{x})|}{|\boldsymbol{y}-\boldsymbol{x}|} \\ &\leqslant |\boldsymbol{v}(\boldsymbol{x}+\alpha(\boldsymbol{y}-\boldsymbol{x}))-\boldsymbol{v}(\boldsymbol{x})|,\end{aligned}$$

where the Schwarz inequality has been used to estimate the right-hand side. On letting $\boldsymbol{y}\to\boldsymbol{x}$ and using the continuity of $\boldsymbol{v}(\cdot)$ we find that $\boldsymbol{v}(\boldsymbol{x})=\operatorname{grad}\varphi(\boldsymbol{x})$, which is what we wanted to show.

It should perhaps be remarked that if $\boldsymbol{v}(\cdot)$ is the gradient $\operatorname{grad}\varphi(\cdot)$ of a scalar field $\varphi(\cdot)$ and if $\boldsymbol{f}(\cdot)$ is a closed path the line integral of $\boldsymbol{v}(\cdot)$ around $\boldsymbol{f}(\cdot)$ is

$$I(\boldsymbol{v}(\cdot),\boldsymbol{f}(\cdot))=\varphi(\boldsymbol{f}(+\infty))-\varphi(\boldsymbol{f}(-\infty))=0\,,$$

which means that the stated condition on $\boldsymbol{v}(\cdot)$ is not merely sufficient for it to be the gradient of a scalar field but is also necessary.

If $\boldsymbol{M}$ is a constant tensor we may want to know if it is symmetric or not and again it happens that a useful way of characterising the symmetry of $\boldsymbol{M}$ is by means of line integrals around closed paths. Indeed the *tensor* $\boldsymbol{M}$ *is symmetric if and only if*

$$\int_{-\infty}^{+\infty}\dot{\boldsymbol{f}}(t)\cdot\boldsymbol{M}\boldsymbol{f}(t)\,dt\geqslant 0 \tag{1.2.3}$$

for every closed path $\boldsymbol{f}(\cdot)$ in $\mathscr{U}$. To see that this is true we need only note that if the inequality (1.2.3) holds then, as we have just seen, there is a scalar field $\varphi(\cdot)$ whose gradient is

$$\operatorname{grad}\varphi(\boldsymbol{x})=\boldsymbol{M}\boldsymbol{x}\,,$$

everywhere on $\mathscr{U}$. Accordingly the second gradient $\operatorname{grad}^2\varphi(\cdot)$ of $\varphi(\cdot)$ must equal $\boldsymbol{M}$ itself. However, $\operatorname{grad}^2\varphi(\cdot)$ is necessarily symmetric, which means that $\boldsymbol{M}$ must be symmetric. On the other hand, if $\boldsymbol{M}$ is symmetric and $\boldsymbol{f}(\cdot)$ is a path

$$\frac{d}{dt}(\boldsymbol{f}(t)\cdot\boldsymbol{M}\boldsymbol{f}(t))=2\dot{\boldsymbol{f}}(t)\cdot\boldsymbol{M}\boldsymbol{f}(t)$$

from which it follows on integrating that

$$\int_{-\infty}^{+\infty}\dot{\boldsymbol{f}}(t)\cdot\boldsymbol{M}\boldsymbol{f}(t)\,dt=\tfrac{1}{2}\boldsymbol{f}(+\infty)\cdot\boldsymbol{M}\boldsymbol{f}(+\infty)-\tfrac{1}{2}\boldsymbol{f}(-\infty)\cdot\boldsymbol{M}\boldsymbol{f}(+\infty)\,. \tag{1.2.4}$$

The right-hand side of equation (1.2.4) vanishes if $\boldsymbol{f}(\cdot)$ is closed, which confirms that the condition (1.2.3) is necessary as well as sufficient for $\boldsymbol{M}$ to be symmetric.

It is not difficult to see that the two results of this section can be strengthened in the sense that if $\boldsymbol{x}_0$ is some fixed point of $\mathscr{U}$ we need consider only those closed paths in $\mathscr{U}$ which start and end at $\boldsymbol{x}_0$: *the vector field* $\boldsymbol{v}(\cdot)$ *can be expressed as the gradient of a scalar field on* $\mathscr{U}$ *if and only if* $I(\boldsymbol{v}(\cdot),\boldsymbol{f}(\cdot))\geqslant 0$ *for every closed path* $\boldsymbol{f}(\cdot)$ *with* $\boldsymbol{f}(-\infty)=\boldsymbol{f}(+\infty)=\boldsymbol{x}_0$, *and the tensor* $\boldsymbol{M}$ *is symmetric if and only if*

$$\int_{-\infty}^{+\infty}\dot{\boldsymbol{f}}(t)\cdot\boldsymbol{M}\boldsymbol{f}(t)\,dt\geqslant 0 \tag{1.2.5}$$

for every closed path $\boldsymbol{f}(\cdot)$ *with* $\boldsymbol{f}(-\infty)=\boldsymbol{f}(+\infty)=\boldsymbol{x}_0$.

1.3 Kinematics and the Balance Laws

The kinematics of continuous bodies and the balance laws for mass, momentum, moment of momentum and energy and the various deductions which follow from these laws are all matters which are discussed in detail in the treatise of Truesdell and Toupin[1] and the reader is referred to that work for a full account of them. Our concern here is merely to assemble the few results we need and to introduce the various scalar, vector and tensor fields which describe the mechanical and thermal behaviour of simple materials.

One way of describing the motion of a body is to suppose it placed in a configuration which we adopt as a reference configuration and to label the particles of the body by the positions they occupy in this reference configuration. This means that we can speak of "the particle which is at $\boldsymbol{X}$ in the reference configuration" or, more briefly, of "the particle $\boldsymbol{X}$". The motion of the body is known once we know the position $\boldsymbol{p}(\boldsymbol{X},t)$ occupied by each particle $\boldsymbol{X}$ at each time t. The function $\boldsymbol{p}(\cdot,\cdot)$ cannot be entirely arbitrary; we shall always assume it to be differentiable as many times as we need but, more important, it is restricted by the fact that two different particles $\boldsymbol{X}$ and $\boldsymbol{Y}$ cannot occupy the same position at the same time, in other words $\boldsymbol{p}(\boldsymbol{X},t)\neq\boldsymbol{p}(\boldsymbol{Y},t)$ if $\boldsymbol{X}\neq\boldsymbol{Y}$. We can, of course, always decide to work not with the material coordinates $(\boldsymbol{X},t)$ as independent variables and instead we can regard the equation $\boldsymbol{x}=\boldsymbol{p}(\boldsymbol{X},t)$ as being inverted to give $\boldsymbol{X}$ in terms of $\boldsymbol{x}$ and t and used to express all scalar, vector and tensor fields as functions of the spatial coordinates $(\boldsymbol{x},t)$. In order to distinguish the gradients taken with respect to the variables $\boldsymbol{x}$ and the variables $\boldsymbol{X}$ we write

$$\operatorname{grad}\varphi(\boldsymbol{x},t)=\left(\frac{\partial\varphi}{\partial x_i}(\boldsymbol{x},t)\right), \qquad \operatorname{GRAD}\psi(\boldsymbol{X},t)=\left(\frac{\partial\psi}{\partial X_i}(\boldsymbol{X},t)\right) \qquad (i=1,2,3). \tag{1.3.1}$$

We use a superposed dot to denote the material time derivative:

$$\dot{\psi}(\boldsymbol{X},t)=\frac{\partial\psi}{\partial t}(\boldsymbol{X},t).$$

The velocity and acceleration fields associated with the motion are $\boldsymbol{v}(\cdot,\cdot)=\dot{\boldsymbol{p}}(\cdot,\cdot)$ and $\boldsymbol{a}(\cdot,\cdot)=\ddot{\boldsymbol{p}}(\cdot,\cdot)$ and the material time derivative of a function $\varphi(\cdot,\cdot)$ is given by the usual formula, namely

$$\dot{\varphi}(\boldsymbol{x},t)=\frac{\partial\varphi}{\partial t}(\boldsymbol{x},t)+\boldsymbol{v}(\boldsymbol{x},t)\cdot\operatorname{grad}\varphi(\boldsymbol{x},t). \tag{1.3.2}$$

---

[1] See Truesdell and Toupin [69]. The article [53] of Gurtin and Williams treats the balance law for energy and the Clausius-Duhem inequality axiomatically in the context of rigid heat conductors. See also Williams [73, 74].

The assumptions we have made on the function $\boldsymbol{p}(\cdot,\cdot)$ imply that the determinant of the *deformation gradient tensor*

$$\boldsymbol{F}(\cdot,\cdot)=\mathrm{GRAD}\,\boldsymbol{p}(\cdot,\cdot) \tag{1.3.3}$$

never vanishes. Moreover, if the reference configuration is one which is occupied by the body at some time the tensor $\boldsymbol{F}$ must have a positive determinant and we shall always assume that this is so. The tensor $\boldsymbol{F}$ provides a local description of the effect of the deformation and, as we pointed out in section 1.1, it can be decomposed into the product

$$\boldsymbol{F}=\boldsymbol{R}\boldsymbol{U} \tag{1.3.4}$$

of an orthogonal tensor $\boldsymbol{R}$ with a positive definite and symmetric tensor $\boldsymbol{U}$. In effect equation (1.3.4) asserts that the deformation can be resolved, at least locally, into a pure strain of the material, represented by $\boldsymbol{U}$, followed by a rigid rotation, represented by $\boldsymbol{R}$. Although $\boldsymbol{U}$ does provide a measure of the local strain it is often more convenient to work with the Cauchy-Green tensor

$$\boldsymbol{C}=\boldsymbol{U}^2=\boldsymbol{F}^T\boldsymbol{F} \tag{1.3.5}$$

whose components are rational functions of the components of $\boldsymbol{F}$ whereas the components of $\boldsymbol{U}$ are not.

If ρ is the mass density of the body, if ρ_0 is the mass density in the reference configuration and if $\mathcal{R}$ is the region occupied by a part of the body then, because mass is conserved,

$$\int_{\mathcal{R}_0} \rho_0(\boldsymbol{X})\,dV(\boldsymbol{X})=\int_{\mathcal{R}} \rho(\boldsymbol{x},t)\,dV(\boldsymbol{x}),$$

where $\mathcal{R}_0$ is the region occupied in the reference configuration. On making the change of variable $\boldsymbol{x}=\boldsymbol{p}(\boldsymbol{X},t)$ in the integral on the right-hand side it follows straightaway that ρ_0 and ρ are related by the equation

$$\rho_0=\rho\det\boldsymbol{F}. \tag{1.3.6}$$

If we now take the material time derivative of both sides of equation (1.3.6) and use the identity

$$\operatorname{grad}\boldsymbol{v}=\dot{\boldsymbol{F}}\boldsymbol{F}^{-1} \tag{1.3.7}$$

and equation (1.3.2) we deduce the usual expression, in the spatial description, for balance of mass, namely

$$\frac{\partial\rho}{\partial t}+\operatorname{div}(\rho\boldsymbol{v})=0, \tag{1.3.8}$$

in which $\operatorname{div}(\rho\boldsymbol{v})=\operatorname{trace}\operatorname{grad}(\rho\boldsymbol{v})$ is the spatial divergence of the vector $\rho\boldsymbol{v}$.

In order to set up the balance laws for momentum, moment of momentum and energy we must account for the mechanical and thermal action exerted by one part of the body upon another. We take it that if a part occupies the region $\mathscr{R}$ at time t and if the vector $\boldsymbol{n}(\cdot,\cdot)$ is the unit outward normal to the bounding surface $\partial\mathscr{R}$ of $\mathscr{R}$ the mechanical action of the exterior of the part arises from a contact force

$$\int_{\partial\mathscr{R}} \boldsymbol{t}(\boldsymbol{x},\boldsymbol{n}(\boldsymbol{x},t),t)\,d\mathscr{A}(\boldsymbol{x}),$$

which is computed from a stress vector field $\boldsymbol{t}(\boldsymbol{x},\boldsymbol{n},t)$ per unit area defined on the boundary $\partial\mathscr{R}$, and from a body force

$$\int_{\mathscr{R}} \rho(\boldsymbol{x},t)\,\boldsymbol{b}(\boldsymbol{x},t)\,dV(\boldsymbol{x}),$$

computed from a body force field $\boldsymbol{b}(\boldsymbol{x},t)$ per unit mass defined on $\mathscr{R}$. To take account of the thermal action we assume as well that energy is supplied to $\mathscr{R}$ by heat conduction at a rate

$$\int_{\partial\mathscr{R}} q(\boldsymbol{x},\boldsymbol{n}(\boldsymbol{x},t),t)\,d\mathscr{A}(\boldsymbol{x})$$

where $q(\boldsymbol{x},\boldsymbol{n},t)$ is the scalar heat flux per unit area defined on $\partial\mathscr{R}$ and also that energy is supplied to $\mathscr{R}$ at a rate

$$\int_{\mathscr{R}} \rho(\boldsymbol{x},t)\,r(\boldsymbol{x},t)\,dV(\boldsymbol{x})$$

by a scalar heat supply field $r(\boldsymbol{x},t)$ per unit mass defined on $\mathscr{R}$. It is possible to start with more general assumptions than the ones made here and the reader is referred to the treatise of Truesdell and Toupin [69] for an account of them.

Once these assumptions about the action of the exterior of a part upon the part itself have been made one can set up the usual global balance laws for momentum and moment of momentum and it can then be deduced, by a standard argument, that the balance of momentum implies that the dependence of the stress vector field $\boldsymbol{t}(\boldsymbol{x},\boldsymbol{n},t)$ upon $\boldsymbol{n}$ is

$$\boldsymbol{t}(\boldsymbol{x},\boldsymbol{n},t)=\boldsymbol{T}(\boldsymbol{x},t)\,\boldsymbol{n} \tag{1.3.9}$$

where $\boldsymbol{T}$ is the *stress tensor*. It can also be shown, because of the balance of moment of momentum, that the stress tensor must be symmetric, that is

$$\boldsymbol{T}=\boldsymbol{T}^T \tag{1.3.10}$$

and, having proved the existence of the stress tensor field, the balance of energy tells us that the dependence of the scalar heat flux $q(\boldsymbol{x},\boldsymbol{n},t)$ upon $\boldsymbol{n}$ is

$$q(\boldsymbol{x},\boldsymbol{n},t)=-\boldsymbol{q}(\boldsymbol{x},t)\cdot\boldsymbol{n}, \tag{1.3.11}$$

where $\boldsymbol{q}$ is the *heat conduction vector*. The minus sign occurring on the right-hand side of equation (1.3.11) is conventional: if $\boldsymbol{n}$ is the unit outward normal to the surface $\partial\mathscr{R}$, the scalar product $\boldsymbol{q}\cdot\boldsymbol{n}$ represents the rate of flux of heat energy by conduction from $\mathscr{R}$ to its exterior outwards across the surface $\partial\mathscr{R}$. With the simplifications (1.3.9) and (1.3.11) in the forms of $\boldsymbol{t}$ and q the balance laws for momentum and energy become the statements

$$\int_{\partial\mathscr{R}} \boldsymbol{T}\boldsymbol{n}\,d\mathscr{A} + \int_{\mathscr{R}} \rho\,\boldsymbol{b}\,dV = \int_{\mathscr{R}} \rho\,\boldsymbol{a}\,dV \tag{1.3.12}$$

and

$$\frac{d}{dt}\int_{\mathscr{R}} \rho(e+\tfrac{1}{2}\boldsymbol{v}\cdot\boldsymbol{v})\,dV = \int_{\partial\mathscr{R}} (\boldsymbol{v}\cdot\boldsymbol{T}\boldsymbol{n}-\boldsymbol{q}\cdot\boldsymbol{n})\,d\mathscr{A} + \int_{\mathscr{R}} \rho(\boldsymbol{v}\cdot\boldsymbol{b}+r)\,dV, \tag{1.3.13}$$

where the scalar field $e(\boldsymbol{x},t)$ is the *internal energy* of the body per unit mass. A straightforward application of the divergence theorem to the region $\mathscr{R}$ now tells us that these two laws are together equivalent to the local equations

$$\operatorname{div}\boldsymbol{T} + \rho\,\boldsymbol{b} = \rho\,\boldsymbol{a} \tag{1.3.14}$$

and

$$\rho\dot{e} = \boldsymbol{T}\cdot\operatorname{grad}\boldsymbol{v} - \operatorname{div}\boldsymbol{q} + \rho r\,. \tag{1.3.15}$$

The local equations (1.3.6), (1.3.10), (1.3.14) and (1.3.15) are deductions from the global balance laws for mass, momentum, moment of momentum and energy; conversely when they are taken together they also ensure that the balance laws hold.

Up to this point the idea of temperature has not been needed. It will be assumed from here onwards that it is both possible and natural to speak of the *absolute temperature* field $\theta(\cdot,\cdot)$ of the body in all the various processes contemplated for the body, whether these processes are "close to equilibrium" in some sense or not. For us the adjective "absolute" will mean, among other things, that the values $\theta(\boldsymbol{x},t)$ of the temperature field are always strictly positive numbers.

We have now introduced all the scalar, vector and tensor fields required for the theory presented in Chapters 2, 3 and 4. However, the theory described in Chapter 5 and based on the Clausius-Duhem inequality requires that we introduce from the outset one additional scalar field which is denoted by the symbol $\eta(\cdot,\cdot)$ and is interpreted as the *entropy* per unit mass of the body. This means that the entropy of a part of the body occupying the region $\mathscr{R}$ at time t is

$$\int_{\mathscr{R}} \rho(\boldsymbol{x},t)\eta(\boldsymbol{x},t)\,dV(\boldsymbol{x})\,.$$

1.4 Simple Materials with Memory

As we remarked above the local equations (1.3.6), (1.3.10), (1.3.14) and (1.3.15) are equivalent to the global balance laws for mass, momentum, moment of momentum and energy. By themselves they do not suffice to determine the whole mechanical and thermal behaviour of the body and they must be supplemented by constitutive relations appropriate to the material from which the body is made.

The most general materials we shall envisage for the theory of Chapters 2, 3 and 4 are those in which the stress tensor $\boldsymbol{T}(\boldsymbol{X},t)$, the heat flux vector $\boldsymbol{q}(\boldsymbol{X},t)$ and the internal energy $e(\boldsymbol{X},t)$ at a particle $\boldsymbol{X}$ of the body at time t are determined by the values $\boldsymbol{F}(\boldsymbol{X},s)$, $\theta(\boldsymbol{X},s)$, $\boldsymbol{g}(\boldsymbol{X},s)$ taken by the deformation gradient tensor $\boldsymbol{F}(\cdot,\cdot)$, the absolute temperature $\theta(\cdot,\cdot)$ and its spatial gradient $\boldsymbol{g}(\cdot,\cdot)=\operatorname{grad}\theta(\cdot,\cdot)$ at the particle $\boldsymbol{X}$ at all times $s\leqslant t$ prior to the time t. More explicitly, in these materials the stress, the heat flux and the internal energy are given by constitutive relations

$$\boldsymbol{T}(\boldsymbol{X},t)=\underset{s=-\infty}{\overset{t}{\boldsymbol{T}}}(\boldsymbol{F}(\boldsymbol{X},s),\theta(\boldsymbol{X},s),\boldsymbol{g}(\boldsymbol{X},s);\boldsymbol{X}), \tag{1.4.1}$$

$$\boldsymbol{q}(\boldsymbol{X},t)=\underset{s=-\infty}{\overset{t}{\boldsymbol{q}}}(\boldsymbol{F}(\boldsymbol{X},s),\theta(\boldsymbol{X},s),\boldsymbol{g}(\boldsymbol{X},s);\boldsymbol{X}), \tag{1.4.2}$$

$$e(\boldsymbol{X},t)=\underset{s=-\infty}{\overset{t}{e}}(\boldsymbol{F}(\boldsymbol{X},s),\theta(\boldsymbol{X},s),\boldsymbol{g}(\boldsymbol{X},s);\boldsymbol{X}) \tag{1.4.3}$$

in which the response functionals $\underset{s=-\infty}{\overset{t}{\boldsymbol{T}}}$, $\underset{s=-\infty}{\overset{t}{\boldsymbol{q}}}$, $\underset{s=-\infty}{\overset{t}{e}}$ for the stress, heat flux and internal energy are characteristic of the material. The theory of Chapter 5, which uses the Clausius-Duhem inequality, requires us to supplement these relations by an additional relation

$$\eta(\boldsymbol{X},t)=\underset{s=-\infty}{\overset{t}{\eta}}(\boldsymbol{F}(\boldsymbol{X},s),\theta(\boldsymbol{X},s),\boldsymbol{g}(\boldsymbol{X},s);\boldsymbol{X}) \tag{1.4.4}$$

of the same kind for the entropy $\eta(\boldsymbol{X},t)$.

Each of the functionals $\underset{s=-\infty}{\overset{t}{\boldsymbol{T}}}$, $\underset{s=-\infty}{\overset{t}{\boldsymbol{q}}}$, $\underset{s=-\infty}{\overset{t}{e}}$ and $\underset{s=-\infty}{\overset{t}{\eta}}$ is allowed to depend explicitly on the particle $\boldsymbol{X}$, that is to say we allow the possibility that the body is made from an inhomogeneous material whose properties vary from particle to particle, but to save writing we shall usually suppress the dependence upon $\boldsymbol{X}$ and write the constitutive relations in the shorter form

$$\boldsymbol{T}(t)=\underset{s=-\infty}{\overset{t}{\boldsymbol{T}}}(\boldsymbol{F}(s),\theta(s),\boldsymbol{g}(s))=\underset{-\infty}{\overset{t}{\boldsymbol{T}}}(\boldsymbol{F}(\cdot),\theta(\cdot),\boldsymbol{g}(\cdot)), \tag{1.4.5}$$

$$\boldsymbol{q}(t)=\underset{s=-\infty}{\overset{t}{\boldsymbol{q}}}(\boldsymbol{F}(s),\theta(s),\boldsymbol{g}(s))=\underset{-\infty}{\overset{t}{\boldsymbol{q}}}(\boldsymbol{F}(\cdot),\theta(\cdot),\boldsymbol{g}(\cdot)), \tag{1.4.6}$$

$$e(t)=\underset{s=-\infty}{\overset{t}{e}}(\boldsymbol{F}(s),\theta(s),\boldsymbol{g}(s))=\underset{-\infty}{\overset{t}{e}}(\boldsymbol{F}(\cdot),\theta(\cdot),\boldsymbol{g}(\cdot)), \tag{1.4.7}$$

and, where appropriate,

$$\eta(t) = \underset{s=-\infty}{\overset{t}{\eta}}(\boldsymbol{F}(s), \theta(s), \boldsymbol{g}(s)) = \underset{-\infty}{\overset{t}{\eta}}(\boldsymbol{F}(\cdot), \theta(\cdot), \boldsymbol{g}(\cdot)). \tag{1.4.8}$$

Clearly these relations incorporate the idea that the materials have a memory for the past; in some of the arguments given later on we shall need to assume that the memory fades in time in the sense that the values $\boldsymbol{F}(s)$, $\theta(s)$ and $\boldsymbol{g}(s)$ taken in the recent past, the values with s close to t, have greater effect on the values of $\boldsymbol{T}(t)$, $\boldsymbol{q}(t)$ and $e(t)$ (and in the second theory $\eta(t)$ as well) at time t than do the values taken in the remote past. There are several ways of giving fading memory a precise meaning and we shall consider this matter in detail later on. The materials introduced here are *simple materials with memory*; they are "simple" in that the constitutive relations involve only the first gradient of the motion, through $\boldsymbol{F}(\cdot)$, and only the first gradient of the temperature field $\theta(\cdot)$, through $\boldsymbol{g}(\cdot)$.

If they are to be physically realistic the response functionals $\underset{-\infty}{\overset{t}{\boldsymbol{T}}}$, $\underset{-\infty}{\overset{t}{\boldsymbol{q}}}$, $\underset{-\infty}{\overset{t}{e}}$, and $\underset{-\infty}{\overset{t}{\eta}}$ cannot depend on the values taken by $\boldsymbol{F}(\cdot)$, $\theta(\cdot)$ and $\boldsymbol{g}(\cdot)$ in an entirely arbitrary fashion and indeed they must be invariant under rigid body motions in the following sense. Suppose that the body undergoes a process in which the motion is $\boldsymbol{p}(\cdot,\cdot)$ and the temperature field is $\theta(\cdot,\cdot)$. Then the gradients $\boldsymbol{F} = \mathrm{GRAD}\,\boldsymbol{p}(\cdot,\cdot)$ and $\boldsymbol{g} = \mathrm{grad}\,\theta(\cdot,\cdot)$ can be computed readily and the fields $\boldsymbol{T}(\cdot)$, $\boldsymbol{q}(\cdot)$, $e(\cdot)$ and $\eta(\cdot)$ determined from equations (1.4.5), (1.4.6), (1.4.7) and (1.4.8). Now consider a second process in which the motion $\boldsymbol{p}'(\cdot,\cdot)$ of the body differs from the original motion $\boldsymbol{p}(\cdot,\cdot)$ at each instant of time only by a rigid motion, this means that

$$\boldsymbol{p}'(\boldsymbol{X}, s) = \boldsymbol{c}(s) + \boldsymbol{Q}(s)(\boldsymbol{p}(\boldsymbol{X}, s) - \boldsymbol{c}(s)),$$

where $\boldsymbol{c}(s)$ is a variable vector and $\boldsymbol{Q}(s)$ is a variable orthogonal tensor with positive determinant, and in which the material description of the temperature field is unchanged, that is the temperature in the second process is $\theta'(\cdot,\cdot) \equiv \theta(\cdot,\cdot)$. The deformation gradient matrix in the second process is changed to

$$\boldsymbol{F}'(\boldsymbol{X}, s) = \boldsymbol{Q}(s)\boldsymbol{F}(\boldsymbol{X}, s),$$

the spatial temperature gradient is changed to

$$\boldsymbol{g}'(\boldsymbol{X}, s) = \boldsymbol{Q}(s)\boldsymbol{g}(\boldsymbol{X}, s)$$

and the stress, heat flux, internal energy and entropy fields in the new process are

$$\boldsymbol{T}'(t)=\mathop{\boldsymbol{T}}_{s=-\infty}^{t}(\boldsymbol{Q}(s)\boldsymbol{F}(s),\theta(s),\boldsymbol{Q}(s)\boldsymbol{g}(s)), \tag{1.4.9}$$

$$\boldsymbol{q}'(t)=\mathop{\boldsymbol{q}}_{s=-\infty}^{t}(\boldsymbol{Q}(s)\boldsymbol{F}(s),\theta(s),\boldsymbol{Q}(s)\boldsymbol{g}(s)), \tag{1.4.10}$$

$$e'(t)=\mathop{e}_{s=-\infty}^{t}(\boldsymbol{Q}(s)\boldsymbol{F}(s),\theta(s),\boldsymbol{Q}(s)\boldsymbol{g}(s)), \tag{1.4.11}$$

and

$$\eta'(t)=\mathop{\eta}_{s=-\infty}^{t}(\boldsymbol{Q}(s)\boldsymbol{F}(s),\theta(s),\boldsymbol{Q}(s)\boldsymbol{g}(s)). \tag{1.4.12}$$

The components of the tensor $\boldsymbol{T}'(t)$, the components of the vector $\boldsymbol{q}'(t)$ and the values of the scalars $e'(t)$ and $\eta'(t)$ in equations (1.4.9), (1.4.10), (1.4.11) and (1.4.12) can be calculated with respect to the original cartesian coordinate system $\boldsymbol{e}_1$, $\boldsymbol{e}_2$, $\boldsymbol{e}_3$ introduced in section 1.1. However, if we wish to, we can, introduce a rotating coordinate system defined by the vectors $\boldsymbol{e}'_1(s)=\boldsymbol{Q}(s)\boldsymbol{e}_1$, $\boldsymbol{e}'_2(s)=\boldsymbol{Q}(s)\boldsymbol{e}_2$, $\boldsymbol{e}'_3(s)=\boldsymbol{Q}(s)\boldsymbol{e}_3$ and "view" the second process of the body from this non-inertial rotating frame which is moving to keep pace with the rigid rotation superposed on the original motion $\boldsymbol{p}(\cdot,\cdot)$. The invariance requirement is the assumption that the components, with respect to the coordinate system $\boldsymbol{e}'_1(t)$, $\boldsymbol{e}'_2(t)$ and $\boldsymbol{e}'_2(t)$ of the fields $\boldsymbol{T}'(t)$, $\boldsymbol{q}'(t)$, $e'(t)$ and $\eta'(t)$ arising in the second process and determined by equations (1.4.9), (1.4.10), (1.4.11) and (1.4.12) coincide with the components with respect to the fixed coordinate system $\boldsymbol{e}_1, \boldsymbol{e}_2, \boldsymbol{e}_3$ of the fields $\boldsymbol{T}(t)$, $\boldsymbol{q}(t)$, $e(t)$ and $\eta(t)$, arising in the original process and determined by equations (1.4.5), (1.4.6), (1.4.7) and (1.4.8). This means that the response functionals are restricted by the conditions

$$\mathop{\boldsymbol{T}}_{s=-\infty}^{t}(\boldsymbol{Q}(s)\boldsymbol{F}(s),\theta(s),\boldsymbol{Q}(s)\boldsymbol{g}(s))=\boldsymbol{Q}(t)\mathop{\boldsymbol{T}}_{s=-\infty}^{t}(\boldsymbol{F}(s),\theta(s),\boldsymbol{g}(s))\,\boldsymbol{Q}^T(t),$$

$$\mathop{\boldsymbol{q}}_{s=-\infty}^{t}(\boldsymbol{Q}(s)\boldsymbol{F}(s),\theta(s),\boldsymbol{Q}(s)\boldsymbol{g}(s))=\boldsymbol{Q}(t)\mathop{\boldsymbol{q}}_{s=-\infty}^{t}(\boldsymbol{F}(s),\theta(s),\boldsymbol{g}(s)),$$

$$\mathop{e}_{s=-\infty}^{t}(\boldsymbol{Q}(s)\boldsymbol{F}(s),\theta(s),\boldsymbol{Q}(s)\boldsymbol{g}(s))=\mathop{e}_{s=-\infty}^{t}(\boldsymbol{F}(s),\theta(s),\boldsymbol{g}(s)),$$

and

$$\mathop{\eta}_{s=-\infty}^{t}(\boldsymbol{Q}(s)\boldsymbol{F}(s),\theta(s),\boldsymbol{Q}(s)\boldsymbol{g}(s))=\mathop{\eta}_{s=-\infty}^{t}(\boldsymbol{F}(s),\theta(s),\boldsymbol{g}(s))$$

which must hold for each motion $\boldsymbol{p}(\cdot,\cdot)$, each temperature field $\theta(\cdot,\cdot)$ and each time dependent orthogonal tensor $\boldsymbol{Q}(\cdot)$ with positive determinant.

With the aid of the decomposition (1.3.4) of the tensor $\boldsymbol{F}$ it is not difficult to show[1] that the necessary and sufficient conditions for the invariance requirement to hold are that the tensor $\boldsymbol{F}^T\boldsymbol{T}\boldsymbol{F}$, the vector

[1] See, for example, Truesdell and Noll [68].

$\boldsymbol{F}^T \boldsymbol{q}$ and the scalars e and η shall depend on the values taken by $\boldsymbol{F}$ in the past only through the values taken in the past by the tensor $\boldsymbol{C}=\boldsymbol{F}^T \boldsymbol{F}$ and on $\boldsymbol{g}$ only through the values taken by the material temperature gradient $\boldsymbol{F}^T \boldsymbol{g}$ in the past. Equivalently the constitutive relations (1.4.5), (1.4.6) and (1.4.7) can only assume the form

$$\boldsymbol{T}(t)=\boldsymbol{F}(t) \underset{s=-\infty}{\overset{t}{\boldsymbol{T}}} (\boldsymbol{C}(s), \theta(s), \boldsymbol{F}^T(s) \boldsymbol{g}(s)) \boldsymbol{F}^T(t), \tag{1.4.13}$$

$$\boldsymbol{q}(t)=\boldsymbol{F}(t) \underset{s=-\infty}{\overset{t}{\boldsymbol{q}}} (\boldsymbol{C}(s), \theta(s), \boldsymbol{F}^T(s) \boldsymbol{g}(s)), \tag{1.4.14}$$

$$e(t)= \underset{s=-\infty}{\overset{t}{e}} (\boldsymbol{C}(s), \theta(s), \boldsymbol{F}^T(s) \boldsymbol{g}(s)) \tag{1.4.15}$$

and, for a theory based on the Clausius-Duhem inequality, the constitutive relation (1.4.8) must reduce to the equation

$$\eta(t)= \underset{s=-\infty}{\overset{t}{\eta}} (\boldsymbol{C}(s), \theta(s), \boldsymbol{F}^T(s) \boldsymbol{g}(s)). \tag{1.4.16}$$

In principle the response functionals occurring in equations (1.4.13), (1.4.14), (1.4.15) and (1.4.16) should be distinguished from the functionals occurring in equations (1.4.5), (1.4.6), (1.4.7) and (1.4.8) by using a different symbol for them but we shall not do this as it will be clear from the context which functionals are meant.

The forms of the response functionals are restricted further if the material has a particular symmetry, such as isotropy or the symmetry of one of the crystal classes[1]. It turns out that the restrictions introduced by invariance and material symmetry do not play a vital role in either of the thermodynamic theories described in this monograph and consequently it is usually convenient to work not with constitutive relations in which these restrictions are made explicit but to work instead with constitutive relations in the more primitive forms (1.4.5), (1.4.6), (1.4.7) and (1.4.8) on the tacit understanding that the functionals are not entirely arbitrary. It will emerge later that thermodynamic requirements generally restrict the response functionals in a severe way.

There are several important classes of materials which can be regarded as simple materials with memory. To begin with there are *thermoelastic materials* for which the response functionals depend not on the whole past history of $\boldsymbol{F}(\cdot)$, $\theta(\cdot)$ and $\boldsymbol{g}(\cdot)$ but only on their present

---

[1] The reader is referred to the treatise of Truesdell and Noll [68] for an account of the restrictions imposed by material symmetry in certain classes of simple materials.

values, that is to say the relations (1.4.5), (1.4.6), (1.4.7) and (1.4.8) collapse to the equations

$$\boldsymbol{T}(t)=\boldsymbol{T}(\boldsymbol{F}(t),\theta(t),\boldsymbol{g}(t)), \tag{1.4.17}$$

$$\boldsymbol{q}(t)=\boldsymbol{q}(\boldsymbol{F}(t),\theta(t),\boldsymbol{g}(t)), \tag{1.4.18}$$

$$e(t)=e(\boldsymbol{F}(t),\theta(t),\boldsymbol{g}(t)), \tag{1.4.19}$$

$$\eta(t)=\eta(\boldsymbol{F}(t),\theta(t),\boldsymbol{g}(t)) \tag{1.4.20}$$

which, because of the invariance requirement, can only have the form

$$\boldsymbol{T}(t)=\boldsymbol{F}(t)\,\boldsymbol{T}(\boldsymbol{C}(t),\theta(t),\boldsymbol{F}^T(t)\,\boldsymbol{g}(t))\,\boldsymbol{F}^T(t), \tag{1.4.21}$$

$$\boldsymbol{q}(t)=\boldsymbol{F}(t)\,\boldsymbol{q}(\boldsymbol{C}(t),\theta(t),\boldsymbol{F}^T(t)\,\boldsymbol{g}(t)), \tag{1.4.22}$$

$$e(t)=e(\boldsymbol{C}(t),\theta(t),\boldsymbol{F}^T(t)\,\boldsymbol{g}(t)), \tag{1.4.23}$$

$$\eta(t)=\eta(\boldsymbol{C}(t),\theta(t),\boldsymbol{F}^T(t)\,\boldsymbol{g}(t)). \tag{1.4.24}$$

Once again we have not troubled to distinguish the response functions $\boldsymbol{T}(\cdot)$, $\boldsymbol{q}(\cdot)$, $e(\cdot)$ and $\eta(\cdot)$ occurring in equations (1.4.21), (1.4.22), (1.4.23) and (1.4.24) from those occurring in equations (1.4.17), (1.4.18), (1.4.19) and (1.4.20). The restrictions imposed by material symmetry are well-known and they are discussed fully in the treatise of Truesdell and Noll [68].

Among materials in which the past behaviour of $\boldsymbol{F}(\cdot)$, $\theta(\cdot)$ and $\boldsymbol{g}(\cdot)$ does have an important effect on the stress, heat flux, internal energy and entropy there are the materials of integral type and the materials of differential type. The response functionals for materials of the integral type can be approximated by finite sums of multiple integrals. The reader is referred to [68] for an account of this topic. Materials of the integral type may well have a genuine memory for either some finite part of the past history of the fields $\boldsymbol{F}(\cdot)$, $\theta(\cdot)$ and $\boldsymbol{g}(\cdot)$ or even for their entire past history. On the other hand materials of the differential type remember essentially only an arbitrarily small lapse of time into the past; the values of the stress, heat flux, internal energy and entropy fields at time t depend only on the values $\theta(t)$ and $\boldsymbol{g}(t)$ of the temperature and its gradient at that time and on the values of the deformation gradient $\boldsymbol{F}(s)$, for times s near to t. The fact that, when s is near to t, $\boldsymbol{F}(s)$ can be approximated by its Taylor expansion about $s=t$ taken to some suitable order, n say, of powers in $(s-t)$ suggests that we look at constitutive relations of the form

$$\boldsymbol{T}(t)=\boldsymbol{T}(\boldsymbol{F}(t),\boldsymbol{F}^{(1)}(t),\ldots,\boldsymbol{F}^{(n)}(t),\theta(t),\boldsymbol{g}(t)), \tag{1.4.25}$$

$$\boldsymbol{q}(t)=\boldsymbol{q}(\boldsymbol{F}(t),\boldsymbol{F}^{(1)}(t),\ldots,\boldsymbol{F}^{(n)}(t),\theta(t),\boldsymbol{g}(t)), \tag{1.4.26}$$

$$e(t)=\theta(\boldsymbol{F}(t),\boldsymbol{F}^{(1)}(t),\ldots,\boldsymbol{F}^{(n)}(t),\theta(t),\boldsymbol{g}(t)), \tag{1.4.27}$$

$$\eta(t)=\eta(\boldsymbol{F}(t),\boldsymbol{F}^{(1)}(t),\ldots,\boldsymbol{F}^{(n)}(t),\theta(t),\boldsymbol{g}(t)) \tag{1.4.28}$$

where $\boldsymbol{F}^{(k)}(t)$ is the kth derivative $d^k \boldsymbol{F}(t)/dt^k$. These relations define a *material of the differential type* and the number n is called its *complexity.* The explicit restrictions imposed by invariance and by the isotropy of the material are well-known and they too can be found in the treatise of Truesdell and Noll [68].

The *compressible linearly viscous fluid obeying Fourier's law of heat conduction* is a material of the differential type. Its complexity is, of course, $n=1$ and its constitutive relations are

$$\boldsymbol{T}=-p(v,\theta)\boldsymbol{I}+\lambda(v,\theta)(\operatorname{trace}\boldsymbol{D})\boldsymbol{I}+2\mu(v,\theta)\boldsymbol{D}, \tag{1.4.29}$$

$$\boldsymbol{q}=-\kappa(v,\theta)\boldsymbol{g}, \tag{1.4.30}$$

$$e=e(v,\theta), \tag{1.4.31}$$

and, for a theory using the Clausius-Duhem inequality,

$$\eta=\eta(v,\theta), \tag{1.4.32}$$

where $v=1/\rho$ is the *specific volume* of the material, p is the *pressure,* μ is the *shear viscosity,* $\lambda+\frac{2}{3}\mu$ is the *bulk viscosity,* κ is the *thermal conductivity* and where

$$\boldsymbol{D}=\tfrac{1}{2}(\operatorname{grad}\boldsymbol{v}+(\operatorname{grad}\boldsymbol{v})^T) \tag{1.4.33}$$

is the *rate of strain tensor.*

CHAPTER 2

A Theory of Thermodynamics

This Chapter and the next two describe a theory of thermodynamics for simple materials with memory in which entropy enters as a derived concept.

2.1 Processes

The mechanical and thermal behaviour of simple materials is described by seven scalar, vector and tensor fields:—they are the motion $\boldsymbol{p}(\cdot,\cdot)$, the temperature $\theta(\cdot,\cdot)$, the stress tensor $\boldsymbol{T}(\cdot,\cdot)$, the body force vector $\boldsymbol{b}(\cdot,\cdot)$ per unit mass, the heat flux vector $\boldsymbol{q}(\cdot,\cdot)$, the scalar heat supply $r(\cdot,\cdot)$ per unit mass and the internal energy $e(\cdot,\cdot)$ per unit mass. In a physically possible process the local equations (1.3.6), (1.3.10), (1.3.14) and (1.3.15) and the constitutive relations (1.4.5), (1.4.6) and (1.4.7) must hold; if they do we call the collection of seven fields a *process* for the body. We shall suppose that the tensors determined by the stress response functional $\underset{-\infty}{\overset{t}{\boldsymbol{T}}}$ are always symmetric, so that equation (1.3.10) holds, and that the mass density field $\rho(\cdot,\cdot)$ is computed using the local equation (1.3.6) for the balance of mass. It is clear that in order to specify a process for the body all we need do is to specify the motion $\boldsymbol{p}(\cdot,\cdot)$ and the temperature field $\theta(\cdot,\cdot)$ for then the gradients $\boldsymbol{F}=\mathrm{G{\scriptsize RAD}}\,\boldsymbol{p}(\cdot,\cdot)$ and

$$\boldsymbol{g}=\operatorname{grad}\theta=(\boldsymbol{F}^T)^{-1}\,\mathrm{G{\scriptsize RAD}}\,\theta(\cdot,\cdot)$$

can be calculated immediately and the values of $\boldsymbol{T}(\cdot,\cdot)$, $\boldsymbol{q}(\cdot,\cdot)$ and $e(\cdot,\cdot)$ follow from the constitutive relations. If the remaining two fields $\boldsymbol{b}(\cdot,\cdot)$ and $r(\cdot,\cdot)$ are then chosen to satisfy the local equations (1.3.14) and (1.3.15) we produce a process for the body. Furthermore any process can be obtained in this way starting from the motion and the temperature field for the process. The fields $\boldsymbol{b}(\cdot,\cdot)$ and $r(\cdot,\cdot)$ determined by equations (1.3.14) and (1.3.15) are the body force and the heat supply needed to maintain the motion $\boldsymbol{p}(\cdot,\cdot)$ and the temperature field $\theta(\cdot,\cdot)$ and, at least in principle, the motion and the temperature can be varied at will by varying $\boldsymbol{b}(\cdot,\cdot)$ and $r(\cdot,\cdot)$ in a suitable way. The fact that practical difficulties prevent us from varying $\boldsymbol{b}(\cdot,\cdot)$ and $r(\cdot,\cdot)$ arbitrarily in the laboratory, and thereby prevent us from varying $\boldsymbol{p}(\cdot,\cdot)$ and $\theta(\cdot,\cdot)$ arbitrarily, does not affect our argument any more than our inability to produce arbitrary forces acting on mass points prevents us from calculating, on the basis of Newton's law of motion, the force required to produce a given, but arbitrary, motion of a mass point.

To guarantee the validity of the arguments given later on we choose to make two assumptions about the response functional $\underset{-\infty}{\overset{t}{e}}(\boldsymbol{F}(\cdot),\theta(\cdot),\boldsymbol{g}(\cdot))$ for the internal energy; they are not the most general assumptions which might be made but they suffice for our present purpose. It will be assumed that the values $e(t)$ of the response functional are always strictly positive and that the functional can be inverted in the sense that if $\hat{\boldsymbol{F}}(\cdot)$ is any continuous tensor function, with $\det \hat{\boldsymbol{F}}(t)>0$ for every t, and if $\hat{\boldsymbol{g}}(\cdot)$ is a given continuous vector function then corresponding to each scalar function $\hat{e}(\cdot)$ there is just one positive and continuous scalar function $\hat{\theta}(\cdot)$ satisfying the equation

$$\hat{e}(t)=\underset{-\infty}{\overset{t}{e}}(\hat{\boldsymbol{F}}(\cdot),\hat{\theta}(\cdot),\hat{\boldsymbol{g}}(\cdot)), \qquad -\infty<t<+\infty. \tag{2.1.1}$$

The significance of these assumptions is that if $\hat{\boldsymbol{X}}$ is any particle of the body we can always find processes in which the deformation gradient, the internal energy and the spatial temperature gradient at $\hat{\boldsymbol{X}}$ have the preassigned values

$$\boldsymbol{F}(\hat{\boldsymbol{X}},t)=\hat{\boldsymbol{F}}(t), \quad e(\hat{\boldsymbol{X}},t)=\hat{e}(t), \quad \boldsymbol{g}(\hat{\boldsymbol{X}},t)=\hat{\boldsymbol{g}}(t), \qquad -\infty<t<+\infty \tag{2.1.2}$$

where $\hat{\boldsymbol{F}}(\cdot)$, $\hat{e}(\cdot)$ and $\hat{\boldsymbol{g}}(\cdot)$ are the functions introduced above. To verify this we need only consider, for example, the process generated by the homogeneous motion

$$\boldsymbol{p}(\hat{\boldsymbol{X}},t)=\boldsymbol{c}(t)+\hat{\boldsymbol{F}}(t)(\boldsymbol{X}-\hat{\boldsymbol{X}}),$$

in which $\boldsymbol{c}(\cdot)$ is a vector function, and by the temperature field

$$\theta(\boldsymbol{X},t)=\hat{\theta}(t)+\tfrac{1}{2}\hat{\theta}(t)\tan^{-1}\left(\frac{2}{\hat{\theta}(t)}\hat{\boldsymbol{F}}^T(t)\hat{\boldsymbol{g}}(t)\cdot(\boldsymbol{X}-\hat{\boldsymbol{X}})\right),$$

where $\hat{\theta}(t)$ is the solution of the functional equation (2.1.1). It is a straightforward matter to check that the required conditions (2.1.2) hold at the particle $\hat{\boldsymbol{X}}$ and that, for each particle $\boldsymbol{X}$ and for each time t, $\det \boldsymbol{F}(\boldsymbol{X},t)>0$ and $\theta(\boldsymbol{X},t)>0$.

Because of our assumptions on the response functional $\underset{-\infty}{\overset{t}{e}}$ the constitutive relations (1.4.5), (1.4.6) and (1.4.7) can be inverted to a form in which $\boldsymbol{F}, e$ and $\boldsymbol{g}$ are the independent variables:—

$$\boldsymbol{T}(t) = \underset{-\infty}{\overset{t}{\bar{\boldsymbol{T}}}}\,(\boldsymbol{F}(\cdot), e(\cdot), \boldsymbol{g}(\cdot)), \tag{2.1.3}$$

$$\boldsymbol{q}(t) = \underset{-\infty}{\overset{t}{\bar{\boldsymbol{q}}}}\,(\boldsymbol{F}(\cdot), e(\cdot), \boldsymbol{g}(\cdot)), \tag{2.1.4}$$

$$\theta(t) = \underset{-\infty}{\overset{t}{\bar{\theta}}}\,(\boldsymbol{F}(\cdot), e(\cdot), \boldsymbol{g}(\cdot)). \tag{2.1.5}$$

The response functionals $\underset{-\infty}{\overset{t}{\bar{\boldsymbol{T}}}}$ and $\underset{-\infty}{\overset{t}{\bar{\boldsymbol{q}}}}$ must be carefully distinguished from the functionals $\underset{-\infty}{\overset{t}{\boldsymbol{T}}}$ and $\underset{-\infty}{\overset{t}{\boldsymbol{q}}}$ appearing in equations (1.4.5) and (1.4.6).

2.2 The Thermodynamic Inequality

The starting point for our investigation of thermodynamics is an inequality which will be assumed to hold in certain special processes of the body. The processes are those which start from equilibrium and which are cyclic. More precisely, if $\boldsymbol{X}$ is a particle of the body, we say that a process *starts from equilibrium* at $\boldsymbol{X}$ if the deformation gradient and the internal energy at $\boldsymbol{X}$ are constant and the temperature gradient vanishes at all times before some time t_0, that is if

$$\boldsymbol{F}(\boldsymbol{X},t) \equiv \boldsymbol{F}_0, \qquad e(\boldsymbol{X},t) \equiv e_0 \tag{2.2.1}$$

and

$$\boldsymbol{g}(\boldsymbol{X},t) \equiv \boldsymbol{0} \tag{2.2.2}$$

for all times $t \leqslant t_0$, where $\boldsymbol{F}_0$ is a constant tensor with positive determinant and e_0 is a constant positive scalar. We say, in addition, that the process is *cyclic* at $\boldsymbol{X}$ if the deformation gradient and the internal energy at $\boldsymbol{X}$ return to their initial constant values and remain constant after some later time t_1, that is if the equations (2.2.1) also hold for all times $t \geqslant t_1$. We do not require the temperature gradient $\boldsymbol{g}(\boldsymbol{X},t)$ to vanish after the time t_1.

The fundamental thermodynamic assumption is that *in any cyclic process which starts from equilibrium at the particle* $\boldsymbol{X}$ *the inequality*

$$\int_{t_0}^{t_1} \left\{ -\frac{1}{\rho} \operatorname{div}\left(\frac{1}{\theta}\boldsymbol{q}\right) + \frac{1}{\theta} r \right\}\bigg|_{\boldsymbol{X}} dt \leqslant 0 \tag{2.2.3}$$

holds.

Here the integrand is evaluated at the particle $\boldsymbol{X}$ and the times t_0 and t_1 are as in the definition of a cyclic process starting from equilibrium, just given.

For the present we make no attempt to motivate the inequality (2.2.3) nor shall we attempt to derive it from a statement which is logically more primitive. Our task will be to examine the consequences of the inequality and to show that it produces results of the kind we expect of a thermodynamic theory. Later on we shall see that for extensive classes of materials, but not for all materials, the inequality can be proved to be a consequence of the Clausius-Duhem inequality.

The heat supply $r(\cdot,\cdot)$ needed to maintain the cyclic process is given by the local equation (1.3.15) and if we substitute for $r(\cdot,\cdot)$ from (1.3.15) into the inequality (2.2.3) and use the fact that $\operatorname{grad}\boldsymbol{v}=\dot{\boldsymbol{F}}\boldsymbol{F}^{-1}$ we find that the inequality is equivalent to the inequality

$$\int_{t_0}^{t_1}\left\{\frac{1}{\theta}(\dot{e}-\boldsymbol{S}\cdot\dot{\boldsymbol{F}})+\frac{1}{\rho\theta^2}\boldsymbol{q}\cdot\boldsymbol{g}\right\}dt\leqslant 0 \tag{2.2.4}$$

where the tensor $\boldsymbol{S}$ is the *Piola-Kirchhoff stress tensor* which is related to the stress tensor $\boldsymbol{T}$ by the formulae

$$\boldsymbol{S}=\frac{1}{\rho}\boldsymbol{T}(\boldsymbol{F}^T)^{-1}=\frac{1}{\rho_0}(\det\boldsymbol{F})\,\boldsymbol{T}(\boldsymbol{F}^T)^{-1}\,. \tag{2.2.5}$$

The symmetry of $\boldsymbol{T}$ is equivalent to the condition

$$\boldsymbol{S}\boldsymbol{F}^T=\boldsymbol{F}\boldsymbol{S}^T \tag{2.2.6}$$

on $\boldsymbol{S}$. Because $\boldsymbol{T}$ is given by either of the constitutive relations (1.4.5) or (2.1.3), $\boldsymbol{S}$ too is given by constitutive relations of the same kind:

$$\boldsymbol{S}=\mathop{\boldsymbol{S}}_{-\infty}^{t}(\boldsymbol{F}(\cdot),\theta(\cdot),\boldsymbol{g}(\cdot))=\mathop{\bar{\boldsymbol{S}}}_{-\infty}^{t}(\boldsymbol{F}(\cdot),e(\cdot),\boldsymbol{g}(\cdot))\,. \tag{2.2.7}$$

2.3 Heat Conduction Inequalities

A number of heat conduction inequalities can be deduced immediately from the thermodynamic inequality (2.2.3), or from the equivalent inequality (2.2.4). To do so let $\boldsymbol{F}_0$ be any constant tensor with a positive determinant, let e_0 be any constant positive scalar and let $\boldsymbol{g}(\cdot)$ be any continuous vector function with $\boldsymbol{g}(t)=0$ for every $t\leqslant 0$. As we have seen already the assumptions we have made about the response functional $\mathop{e}_{-\infty}^{t}$ for the internal energy mean that there is a process in which the deformation gradient and the internal energy, at a given particle, have the constant values $\boldsymbol{F}_0$ and e_0 and in which the spatial temperature gradient is $\boldsymbol{g}(\cdot)$. Clearly this process is a cyclic process starting from equilibrium and, indeed, we can take $t_0=0$ in the inequalities (2.2.3) and (2.2.4) and we can take t_1 to be any $u>0$. In this process the derivatives $\dot{\boldsymbol{F}}(\cdot)$ and $\dot{e}(\cdot)$ vanish identically and the inequality (2.2.4) reduces to the inequality

$$\int_0^u \frac{1}{\rho(t)\theta(t)^2}\boldsymbol{q}(t)\cdot\boldsymbol{g}(t)\,dt \leqslant 0,$$

which, since the mass density $\rho=\rho_0/\det \boldsymbol{F}_0$ is constant, can be written more explicitly as the inequality

$$\int_0^u \frac{1}{\left\{\overset{t}{\underset{-\infty}{\bar{\theta}}}(\boldsymbol{F}_0,e_0,\boldsymbol{g}(\cdot))\right\}^2}\overset{t}{\underset{-\infty}{\bar{\boldsymbol{q}}}}(\boldsymbol{F}_0,e_0,\boldsymbol{g}(\cdot))\cdot\boldsymbol{g}(t)\,dt \leqslant 0, \tag{2.3.1}$$

valid for every $u>0$ and for every vector function $\boldsymbol{g}(\cdot)$ with $\boldsymbol{g}(t)\equiv\boldsymbol{0}$ for $t\leqslant 0$.

As it stands the inequality (2.3.1) certainly does not imply that

$$\overset{t}{\underset{-\infty}{\bar{\boldsymbol{q}}}}(\boldsymbol{F}_0,e_0,\boldsymbol{g}(\cdot))\cdot\boldsymbol{g}(t)\leqslant 0 \tag{2.3.2}$$

but it does do so if the dependence of the functionals $\overset{t}{\underset{-\infty}{\bar{\boldsymbol{q}}}}$ and $\overset{t}{\underset{-\infty}{\bar{\theta}}}$ on $\boldsymbol{g}(\cdot)$ is of a suitable kind. The dependence is suitable, for example, in the case of the compressible linearly viscous fluid obeying Fourier's law whose constitutive relations are (1.4.29), (1.4.30), and (1.4.31). For this material the inequality (2.3.1) reduces to the statement that for every $u>0$

$$-\int_0^u \frac{1}{\bar{\theta}(v_0,e_0)^2}\bar{\kappa}(v_0,e_0)\,\boldsymbol{g}(t)\cdot\boldsymbol{g}(t)\,dt \leqslant 0,$$

where v_0 is the constant specific volume of the material in the process. It follows that *the thermal conductivity κ is never negative*

$$\kappa \geqslant 0, \tag{2.3.3}$$

and that the *heat conduction inequality*

$$\boldsymbol{q}\cdot\boldsymbol{g}\leqslant 0, \tag{2.3.4}$$

which says that heat flows from higher to lower temperatures, must hold.

Similar, but more general, results can be derived for the thermoelastic material defined by the relations (1.4.17), (1.4.18) and (1.4.19). In this case the inequality (2.3.1) states that

$$\int_0^u \frac{1}{\{\bar{\theta}(\boldsymbol{F}_0, e_0, \boldsymbol{g}(t))\}^2} \bar{\boldsymbol{q}}(\boldsymbol{F}_0, e_0, \boldsymbol{g}(t)) \cdot \boldsymbol{g}(t)\, dt \leqslant 0\,, \tag{2.3.5}$$

where $\bar{q}(\cdot,\cdot,\cdot)$ and $\bar{\theta}(\cdot,\cdot,\cdot)$ are the response functions for the heat flux and the temperature. If $\boldsymbol{g}_0$ is any constant vector we can define a sequence of vector functions $\boldsymbol{g}_n(\cdot)$ $(n=1,2,3,\ldots)$ by setting $\boldsymbol{g}_n(t)=\mathbf{0}$ for $t\leqslant 0$, $\boldsymbol{g}_n(t)=nt\boldsymbol{g}_0$ for $0\leqslant t\leqslant 1/n$ and $\boldsymbol{g}_n(t)=\boldsymbol{g}_0$ for $t\geqslant 1/n$. As $n\to+\infty$ the sequence of continuous functions $\boldsymbol{g}_n(\cdot)$ approximates a discontinuous step function whose values are $\mathbf{0}$ on $t\leqslant 0$ and $\boldsymbol{g}_0$ on $t>0$. If we replace $\boldsymbol{g}(\cdot)$ by $\boldsymbol{g}_n(\cdot)$ in the inequality (2.3.5) and then let $n\to+\infty$ it follows, if the functions $\bar{\boldsymbol{q}}(\cdot,\cdot,\cdot)$ and $\bar{\theta}(\cdot,\cdot,\cdot)$ are continuous, that for every $u>0$

$$\begin{aligned} 0 &\geqslant \int_0^u \frac{1}{\{\bar{\theta}(\boldsymbol{F}_0, e_0, \boldsymbol{g}_0)\}^2} \bar{\boldsymbol{q}}(\boldsymbol{F}_0, e_0, \boldsymbol{g}_0) \cdot \boldsymbol{g}_0\, dt \\ &= \frac{u}{\{\bar{\theta}(\boldsymbol{F}_0, e_0, \boldsymbol{g}_0\}^2} \bar{\boldsymbol{q}}(\boldsymbol{F}_0, e_0, \boldsymbol{g}_0) \cdot \boldsymbol{g}_0\,. \end{aligned}$$

Accordingly the *heat conduction inequality*

$$\bar{\boldsymbol{q}}(\boldsymbol{F}_0, e_0, \boldsymbol{g}_0) \cdot \boldsymbol{g}_0 \leqslant 0 \tag{2.3.6}$$

holds for every vector $\boldsymbol{g}_0$.

Interesting results follow from this heat conduction inequality[1]. Let us introduce the *conductivity tensor*

$$\boldsymbol{K}(\boldsymbol{F}_0, e_0) = -\operatorname{grad}_{\boldsymbol{g}} \bar{\boldsymbol{q}}|_{\boldsymbol{g}=\mathbf{0}} = -\left(\frac{\partial \bar{q}_i}{\partial g_j}(\boldsymbol{F}_0, e_0, \mathbf{0})\right) \qquad (i,j=1,2,3) \tag{2.3.7}$$

evaluated at zero temperature gradient. The heat conduction inequality tells us that the scalar function $\bar{\boldsymbol{q}}(\boldsymbol{F}_0, e_0, \boldsymbol{g}) \cdot \boldsymbol{g}$ of the vector $\boldsymbol{g}$ has a maximum at $\boldsymbol{g}=\mathbf{0}$. Accordingly at $\boldsymbol{g}=\mathbf{0}$ its first gradient

$$\operatorname{grad}_{\boldsymbol{g}} \{\bar{\boldsymbol{q}}(\boldsymbol{F}_0, e_0, \boldsymbol{g}) \cdot \boldsymbol{g}\} = \bar{\boldsymbol{q}}(\boldsymbol{F}_0, e_0, \boldsymbol{g}) + (\operatorname{grad}_{\boldsymbol{g}} \bar{\boldsymbol{q}}(\boldsymbol{F}_0, e_0, \boldsymbol{g}))^T \boldsymbol{g}$$

vanishes, that is

$$\bar{\boldsymbol{q}}(\boldsymbol{F}_0, e_0, \mathbf{0}) = \mathbf{0}, \tag{2.3.8}$$

[1] The results derived in this section from the heat conduction inequality (2.3.6) are due to Pipkin and Rivlin [63]. We are tacitly assuming that the function $\bar{\boldsymbol{q}}(\cdot,\cdot,\cdot)$ is twice differentiable.

and its second gradient must be a negative semi-definite tensor, which means that

$$\boldsymbol{K}(\boldsymbol{F}_0, e_0) + \boldsymbol{K}^T(\boldsymbol{F}_0, e_0)$$

is a positive semi-definite tensor and hence that the conductivity tensor $\boldsymbol{K}(\boldsymbol{F}_0, e_0)$ is itself positive semi-definite.

Equation (2.3.8) asserts that *it is not possible to produce a flow of heat in a thermoelastic material at zero temperature gradient no matter what strain or temperature fields are applied to it; in other words, the thermodynamic inequality rules out piezo-caloric effects in a thermoelastic material. The positive semi-definiteness of the conductivity tensor generalises the restriction* (2.3.3) *on the scalar conductivity of a linearly viscous fluid.* Because of (2.3.8) the Taylor expansion of $\bar{\boldsymbol{q}}(\cdot,\cdot,\cdot)$ about $\boldsymbol{g}=\boldsymbol{0}$ is

$$\bar{\boldsymbol{q}}(\boldsymbol{F}, e, \boldsymbol{g}) = -\boldsymbol{K}(\boldsymbol{F}, e)\boldsymbol{g} + o(\boldsymbol{g}), \tag{2.3.9}$$

where the remainder term $o(\boldsymbol{g})$ has the property that $o(\boldsymbol{g})/|\boldsymbol{g}| \to 0$ as $\boldsymbol{g} \to \boldsymbol{0}$. In this sense *Fourier's law*

$$\bar{\boldsymbol{q}}(\boldsymbol{F}, e, \boldsymbol{g}) = -\boldsymbol{K}(\boldsymbol{F}, e)\boldsymbol{g}$$

holds approximately when the temperature gradient is small.

It does not follow from the heat conduction inequality (2.3.6) that the conductivity tensor is symmetric—a result which is sometimes claimed to be a special case of the Onsager reciprocal relations. The reader is referred to Chapter 7 of Truesdell [67] for a full discussion of this matter. A thermodynamic axiom which does imply the symmetry of the conductivity tensor for rigid heat conductors can be found in [40].

2.4 The Conversion of Heat into Mechanical Work

The study of thermodynamics began as an attempt to understand the principles governing the working of heat engines. In essence a heat engine is a device for converting heat energy into mechanical work by taking a material, the working substance of the engine, around a cyclic process starting from equilibrium. In practice the temperature in the process is always confined between two working temperatures characteristic of the engine; the upper working temperature, which is the temperature of the boiler, and the lower working temperature, which is the temperature of the condenser.

In this section we derive from the thermodynamic inequality (2.2.4) and the local equation (1.3.15) the classical formula for the efficiency of a cyclic process starting from equilibrium in terms of upper and lower bounds on the temperature in the process. No special assumptions are required about the process itself, which certainly need not be a Carnot cycle, that is a cycle in which heat is absorbed only at one working temperature and emitted only at the other working temperature. We shall, however, make an assumption about the material; the assumption is that *the stress and the internal energy are independent of the temperature gradient* so that the constitutive relations (1.4.5) and (1.4.7) reduce to

$$\boldsymbol{T}(t) = \overset{t}{\underset{-\infty}{\underline{\boldsymbol{T}}}}\,(\boldsymbol{F}(\cdot), \theta(\cdot)) \tag{2.4.1}$$

and

$$e(t) = \overset{t}{\underset{-\infty}{\underline{e}}}\,(\boldsymbol{F}(\cdot), \theta(\cdot)) \tag{2.4.2}$$

and the inverted relations (2.1.3), (2.1.5) to

$$\boldsymbol{T}(t) = \overset{t}{\underset{-\infty}{\underline{\bar{\boldsymbol{T}}}}}\,(\boldsymbol{F}(\cdot), e(\cdot)) \tag{2.4.3}$$

and

$$\theta(t) = \overset{t}{\underset{-\infty}{\underline{\bar{\theta}}}}\,(\boldsymbol{F}(\cdot), e(\cdot))\,. \tag{2.4.4}$$

The Piola-Kirchhoff stress, which was defined in (2.2.5), must also be independent of the temperature gradient:

$$\boldsymbol{S}(t) = \overset{t}{\underset{-\infty}{\underline{\bar{\boldsymbol{S}}}}}\,(\boldsymbol{F}(\cdot), e(\cdot)) \tag{2.4.5}$$

Suppose that $\boldsymbol{F}(\cdot)$ is any tensor function with positive determinant, that $e(\cdot)$ is any positive scalar function and that $\boldsymbol{F}(t) \equiv \boldsymbol{F}_0$ and $e(t) \equiv e_0$ for all times $t \leqslant t_0$ and for all times $t \geqslant t_1$ where $\boldsymbol{F}_0$ is a constant tensor and e_0 is a constant scalar. There is, as we have seen, a process in which, at the time t, the deformation gradient is $\boldsymbol{F}(t)$, the internal energy is $e(t)$ and the temperature gradient vanishes identically. This process is necessarily a cyclic process starting from equilibrium and the inequality (2.2.4) applies to it, which means that the *Clausius inequality*

$$\int_{t_0}^{t_1} \frac{1}{\theta(t)} (\dot{e}(t) - \boldsymbol{S}(t) \cdot \dot{\boldsymbol{F}}(t))\, d t \leqslant 0 \tag{2.4.6}$$

holds.

It is convenient to write the equation (1.3.15) in the form

$$\dot{e} = \boldsymbol{S} \cdot \dot{\boldsymbol{F}} + h\,, \tag{2.4.7}$$

where the scalar function $h(\cdot)$, which is given by

$$h = -\frac{1}{\rho} \operatorname{div} \boldsymbol{q} + r\,, \tag{2.4.8}$$

and which we choose to call the *total heat supply* per unit mass, is the contribution to the rate of increase of the internal energy density arising from the scalar heat supply $r(\cdot)$ and from heat conduction. The term $\boldsymbol{S}\cdot\dot{\boldsymbol{F}}$ is the mechanical contribution to the rate of increase and we shall call

$$w = -\int_{t_0}^{t_1} \boldsymbol{S}(t)\cdot\dot{\boldsymbol{F}}(t)\,dt \tag{2.4.9}$$

the *mechanical work* done by the material. Because of (2.4.7) the Clausius inequality (2.4.6) can be written as

$$\int_{t_0}^{t_1} \frac{h(t)}{\theta(t)}\,dt \leqslant 0\,. \tag{2.4.10}$$

Heat is absorbed by the material at those times t in the interval $t_0 \leqslant t \leqslant t_1$ at which $h(t) > 0$ and heat is emitted at those times at which $h(t) < 0$. Because $\frac{1}{2}(|h(t)| + h(t)) = h(t)$ if $h(t) > 0$, that is if heat is being absorbed at time t, and $\frac{1}{2}(|h(t)| + h(t)) = 0$ if $h(t) < 0$, that is if heat is being emitted, the *heat absorbed* by the material in the process is

$$h^+ = \tfrac{1}{2}\int_{t_0}^{t_1} (|h(t)| + h(t))\,dt \geqslant 0 \tag{2.4.11}$$

and because $\frac{1}{2}(|h(t)| - h(t)) = 0$ if $h(t) > 0$ and $\frac{1}{2}(|h(t)| - h(t)) = -h(t)$ if $h(t) < 0$ the *heat emitted* by the material is

$$h^- = \tfrac{1}{2}\int_{t_0}^{t_1} (|h(t)| - h(t))\,dt \geqslant 0\,. \tag{2.4.12}$$

Integrating both sides of equation (2.4.7) and using the fact that

$$\int_{t_0}^{t_1} \dot{e}(t)\,dt = e(t_1) - e(t_0) = 0$$

gives the equation

$$-\int_{t_0}^{t_1} \boldsymbol{S}(t)\cdot\dot{\boldsymbol{F}}(t)\,dt = \int_{t_0}^{t_1} h(t)\,dt = \tfrac{1}{2}\int_{t_0}^{t_1} (|h(t)| + h(t))\,dt - \tfrac{1}{2}\int_{t_0}^{t_1} (|h(t)| - h(t))\,dt\,,$$

which means that the mechanical work done by the material is, as we should expect, the difference between the heat absorbed and the heat emitted:

$$w = h^+ - h^-\,. \tag{2.4.13}$$

Now suppose that in the process the absolute temperature $\theta(\cdot)$ lies always between an upper working temperature θ^+ and a lower working temperature θ^-, that is to say, $0<\theta^-\leqslant\theta(t)\leqslant\theta^+$ for all times t in $t_0\leqslant t\leqslant t_1$. The inequality (2.4.10) can be written as

$$\tfrac{1}{2}\int_{t_0}^{t_1}\frac{1}{\theta(t)}(|h(t)|+h(t))\,dt-\tfrac{1}{2}\int_{t_0}^{t_1}\frac{1}{\theta(t)}(|h(t)|-h(t))\,dt\leqslant 0.$$

But, since $|h(t)|+h(t)\geqslant 0$,

$$\tfrac{1}{2}\int_{t_0}^{t_1}\frac{1}{\theta(t)}(|h(t)|+h(t))\,dt\geqslant\tfrac{1}{2}\int_{t_0}^{t_1}\frac{1}{\theta^+}(|h(t)|+h(t))\,dt=\frac{h^+}{\theta^+}$$

and, since $|h(t)|-h(t)\geqslant 0$,

$$\tfrac{1}{2}\int_{t_0}^{t_1}\frac{1}{\theta(t)}(|h(t)|-h(t))\,dt\leqslant\tfrac{1}{2}\int_{t_0}^{t_1}\frac{1}{\theta^-}(|h(t)|-h(t))\,dt=\frac{h^-}{\theta^-}$$

and we deduce that

$$\frac{h^+}{\theta^+}-\frac{h^-}{\theta^-}\leqslant 0$$

or, equivalently, that *the ratio of the heat absorbed to the heat emitted in a cyclic process starting from equilibrium does not exceed the ratio of the upper working temperature to the lower working temperature:*

$$\frac{h^+}{h^-}\leqslant\frac{\theta^+}{\theta^-}. \tag{2.4.14}$$

One consequence is that if $h^+>0$ then $h^->0$, that is to say *it is not possible to perform a cyclic process starting from equilibrium in which a positive amount of heat is absorbed but no heat is emitted.* From the equation (2.4.13) and the inequality (2.4.14) there follows the classical inequality

$$\frac{w}{h^+}\leqslant\frac{\theta^+-\theta^-}{\theta^+}<1, \tag{2.4.15}$$

that is to say *in a cyclic process starting from equilibrium the efficiency w/h^+, which is the ratio of the mechanical work done by the material to the heat absorbed by it, is always strictly less than unity.* The conversion of heat into work in materials with memory is discussed in greater detail in the article [33], to which the reader is referred.

CHAPTER 3

The Construction of the Entropy

The major task for a theory of thermodynamics for simple materials with memory which sets out from the inequality (2.2.3) is to construct the entropy and to justify the use of the name entropy for the functional which is constructed. The possibility that the material may have a genuine memory makes the construction a much more difficult matter than it is in classical thermodynamics, where memory is not taken into account. The entropy is not defined until equation (3.4.4) and before reaching the definition we shall need to construct two auxiliary functions and a functional.

The whole of this Chapter is concerned only with the class of materials considered in section 2.4, *namely those for which the stress and the internal energy are independent of the history of the temperature gradient and thus the stress and the temperature are determined by constitutive relations of the kind* (2.4.3) *and* (2.4.4). In addition the materials must be subject to two restrictions over and above the thermodynamic inequality and the restriction just mentioned. The extra restrictions are that they must have fading memory and they must be well-behaved under the operation of retarding processes, in senses which will be explained later.

As we shall see, the entropy generally turns out to be a genuine functional depending on the history of the deformation gradient and the history of the internal energy and not just on their present values.

3.1 The Clausius Inequality

The starting point for the construction is the Clausius inequality in either of the equivalent forms (2.4.6) and (2.4.10). It will be convenient to cast the inequality into a slightly different form. We choose to regard pairs $(\boldsymbol{A}, a)$, consisting of an arbitrary tensor $\boldsymbol{A}$ and an arbitrary scalar a, as vectors in a vector space whose dimension is ten. The *scalar product* of the vectors $(\boldsymbol{A},a)$ and $(\boldsymbol{B},b)$ is

$$(\boldsymbol{A},a)\cdot(\boldsymbol{B},b)=\boldsymbol{A}\cdot\boldsymbol{B}+ab=\sum_{i,j=1,2,3} A_{ij}B_{ij}+ab \tag{3.1.1}$$

and the *norm* of $(\boldsymbol{A},a)$ is

$$|(\boldsymbol{A},a)|=[(\boldsymbol{A},a)\cdot(\boldsymbol{A},a)]^{\frac{1}{2}} \tag{3.1.2}$$

The independent variables appearing in the constitutive equations (2.4.4) and (2.4.5) for the temperature and the Piola-Kirchhoff stress are the functions $\boldsymbol{F}(\cdot)$ and $e(\cdot)$. Because of the restrictions on these functions the vector $(\boldsymbol{F}(s),e(s))$ is, for every s, confined to a subset $\mathscr{U}$ of the ten dimensional space consisting of vectors $(\boldsymbol{A},a)$ with $\det\boldsymbol{A}>0$ and $a>0$. It can be shown[1] that the subset $\mathscr{U}$ is both open and connected.

The definition of a path has been given already in section 1.2. In the present context a *path* is a pair of continuous and piecewise C^∞ functions $(\boldsymbol{F}(\cdot),e(\cdot))$ whose values are in $\mathscr{U}$ and having the property that there are times t_0 and t_1 and constant pairs $(\boldsymbol{A},a)$ and $(\boldsymbol{B},b)$, called the *initial and final values* of the path, such that $(\boldsymbol{F}(t),e(t))\equiv(\boldsymbol{A},a)$ for all times $t\leqslant t_0$ and $(\boldsymbol{F}(t),e(t))=(\boldsymbol{B},b)$ for all times $t\geqslant t_1$. We shall write $\boldsymbol{F}(-\infty)$ for $\boldsymbol{A}$, $e(-\infty)$ for a, $\boldsymbol{F}(+\infty)$ for $\boldsymbol{B}$ and $e(+\infty)$ for b and we shall say that the path *connects* $(\boldsymbol{A},a)$ to $(\boldsymbol{B},b)$. The path is *closed* if the initial and final values coincide, that is if $(\boldsymbol{F}(-\infty),e(-\infty))=(\boldsymbol{F}(+\infty),e(+\infty))$, which happens only if both the deformation gradient and the internal energy return to their initial values after the time t_1.

If we introduce the *generalised stress*

$$\boldsymbol{\Sigma}=\left(-\frac{1}{\theta}\boldsymbol{S},\frac{1}{\theta}\right) \tag{3.1.3}$$

---

[1] The fact that the determinant $\det\boldsymbol{A}$ depends continuously on the tensor $\boldsymbol{A}$ shows straightaway that $\mathscr{U}$ is open. To prove that it is connected we have to show that any two vectors $(\boldsymbol{A}_0,a_0)$ and $(\boldsymbol{A}_1,a_1)$ in $\mathscr{U}$ can be connected by a path $(\boldsymbol{A}(t),a(t))$, $0\leqslant t\leqslant 1$, with $(\boldsymbol{A}(0),a(0))=(\boldsymbol{A}_0,a_0)$ and with $(\boldsymbol{A}(1),a(1))=(\boldsymbol{A}_1,a_1)$. One way to do this is to decompose the tensors $\boldsymbol{A}_0$ and $\boldsymbol{A}_1$ into the products $\boldsymbol{A}_0=\boldsymbol{Q}_0\boldsymbol{P}_0$ and $\boldsymbol{A}_1=\boldsymbol{Q}_1\boldsymbol{P}_1$ in which $\boldsymbol{Q}_0$ and $\boldsymbol{Q}_1$ are orthogonal tensors with positive determinants and $\boldsymbol{P}_0$ and $\boldsymbol{P}_1$ are positive definite and symmetric tensors. There are skew symmetric tensors $\boldsymbol{N}_0$ and $\boldsymbol{N}_1$ whose exponentials are $\exp\boldsymbol{N}_0=\boldsymbol{Q}_0$ and $\exp\boldsymbol{N}_1=\boldsymbol{Q}_1$, and if we define the orthogonal tensor $\boldsymbol{Q}(t)=\exp\{(1-t)\boldsymbol{N}_0+t\boldsymbol{N}_1\}$ and the positive definite and symmetric tensor $\boldsymbol{P}(t)=(1-t)\boldsymbol{P}_0+t\boldsymbol{P}_1$, $0\leqslant t\leqslant 1$, the path

$$(\boldsymbol{A}(t),a(t))=(\boldsymbol{Q}(t)\boldsymbol{P}(t),(1-t)a_0+ta_1),\qquad 0\leqslant t\leqslant 1\,,$$

in $\mathscr{U}$ connects $(\boldsymbol{A}_0,a_0)$ to $(\boldsymbol{A}_1,a_1)$.

then, because of the constitutive relations (2.4.4) and (2.4.5) for θ and $\boldsymbol{S}$, $\boldsymbol{\Sigma}$ is given by the constitutive relation

$$\boldsymbol{\Sigma}(t)=\underset{s=-\infty}{\overset{t}{\boldsymbol{\Sigma}}}(\boldsymbol{F}(s),e(s))=\underset{-\infty}{\overset{t}{\boldsymbol{\Sigma}}}(\boldsymbol{F}(\cdot),e(\cdot)). \tag{3.1.4}$$

With the aid of the scalar product (3.1.1) the integrand in the Clausius inequality (2.4.6) and the integrand in the equivalent inequality (2.4.10), involving the total heat supply h, can be written as

$$\frac{1}{\theta}(\dot{e}-\boldsymbol{S}\cdot\dot{\boldsymbol{F}})=\frac{h}{\theta}=\boldsymbol{\Sigma}\cdot(\dot{\boldsymbol{F}},\dot{e}).$$

If $(\boldsymbol{F}(\cdot),e(\cdot))$ is any path in $\mathscr{U}$ and u_0 and u_1 are any times whatsoever with $u_0<u_1$ we shall write

$$\underset{u_0}{\overset{u_1}{\mathscr{C}}}(\boldsymbol{F}(\cdot),e(\cdot))=\underset{t=u_0}{\overset{u_1}{\mathscr{C}}}(\boldsymbol{F}(t),e(t))=\int_{u_0}^{u_1}\underset{s=-\infty}{\overset{t}{\boldsymbol{\Sigma}}}(\boldsymbol{F}(s),e(s))\cdot(\dot{\boldsymbol{F}}(t),\dot{e}(t))\,dt. \tag{3.1.5}$$

Of course

$$\underset{u_0}{\overset{u_1}{\mathscr{C}}}(\boldsymbol{F}(\cdot),e(\cdot))=\int_{u_0}^{u_1}\frac{h(t)}{\theta(t)}\,dt=\int_{u_0}^{u_1}\frac{1}{\theta(t)}(\dot{e}(t)-\boldsymbol{S}(t)\cdot\dot{\boldsymbol{F}}(t))\,dt \tag{3.1.6}$$

and the integrands vanish in any interval of time in which $\boldsymbol{F}(\cdot)$ and $e(\cdot)$ are both constant. The integral $\underset{-\infty}{\overset{+\infty}{\mathscr{C}}}(\boldsymbol{F}(\cdot),e(\cdot))$ will be called *the Clausius integral* along the path. It is well-defined because the integration over the whole real line can be replaced by an integration over an appropriate finite interval $t_0\leqslant t\leqslant t_1$.

The inequality (2.4.6) can now be restated in the form: *for any closed path* $(\boldsymbol{F}(\cdot),e(\cdot))$ *in* $\mathscr{U}$ *the Clausius inequality*

$$\underset{-\infty}{\overset{+\infty}{\mathscr{C}}}(\boldsymbol{F}(\cdot),e(\cdot))\leqslant 0 \tag{3.1.7}$$

holds.

The Clausius inequality restricts the response functional $\underset{-\infty}{\overset{t}{\boldsymbol{\Sigma}}}$ for the generalised stress and hence it ultimately restricts the response functionals $\underset{-\infty}{\overset{t}{\boldsymbol{T}}}$ and $\underset{-\infty}{\overset{t}{\theta}}$ for the stress and the temperature. When we investigate these restrictions we are led to construct the entropy.

It will be assumed that the response functional for the generalised stress is invariant under a change of origin of the time scale in the sense that if $(\boldsymbol{F}(\cdot),e(\cdot))$ is any path and if t and t_0 are any times then

$$\underset{s=-\infty}{\overset{t+t_0}{\boldsymbol{\Sigma}}}(\boldsymbol{F}(s-t_0),e(s-t_0))=\underset{s=-\infty}{\overset{t}{\boldsymbol{\Sigma}}}(\boldsymbol{F}(s),e(s)). \tag{3.1.8}$$

3.2 Fading Memory

Because the generalised stress $\Sigma(t)$ on a path at time t usually depends not just on the present values $(\boldsymbol{F}(t), e(t))$ but on all the values $(\boldsymbol{F}(s), e(s))$ taken up to the time t the Clausius integral $\overset{+\infty}{\underset{-\infty}{\mathscr{C}}}(\boldsymbol{F}(\cdot), e(\cdot))$ is not an ordinary line integral of the kind considered in section 1.2. The memory of the material for the past complicates matters considerably. The construction of the entropy relies heavily on the assumption that the response functional $\overset{t}{\underset{s=-\infty}{\Sigma}}$ for the generalised stress exhibits fading memory in an appropriate sense.

Suppose that $(\boldsymbol{F}_1(\cdot), e_1(\cdot))$ and $(\boldsymbol{F}_2(\cdot), e_2(\cdot))$ are two paths in $\mathscr{U}$ for which the final value of the first coincides with the initial value of the second, that is $(\boldsymbol{F}_1(+\infty), e_1(+\infty)) = (\boldsymbol{F}_2(-\infty), e_2(-\infty))$. Then there are times u_1 and u_2 with $(\boldsymbol{F}_1(t), e_1(t)) \equiv (\boldsymbol{F}_1(+\infty), e_1(+\infty))$ for every $t \geqslant u_1$ and $(\boldsymbol{F}_2(t), e_2(t)) \equiv (\boldsymbol{F}_2(-\infty), e_2(-\infty))$ for every $t \leqslant u_2$ and from the two paths we can form a family $(\boldsymbol{F}_\lambda(\cdot), e_\lambda(\cdot))$ of *composite* paths depending on the scalar parameter λ by setting $(\boldsymbol{F}_\lambda(t), e_\lambda(t)) = (\boldsymbol{F}_1(t+\lambda), e_1(t+\lambda))$ for $t \leqslant u_2$ and $(\boldsymbol{F}_\lambda(t), e_\lambda(t)) = (\boldsymbol{F}_2(t), e_2(t))$ for $t > u_2$. Provided λ is large enough, in fact if $\lambda > u_1 - u_2$, $(\boldsymbol{F}_\lambda(\cdot), e_\lambda(\cdot))$ is a path which is obtained by traversing the first path, displaced by λ units of time back into the past, and then traversing the second path. Its initial value is $(\boldsymbol{F}_1(-\infty), e_1(-\infty))$

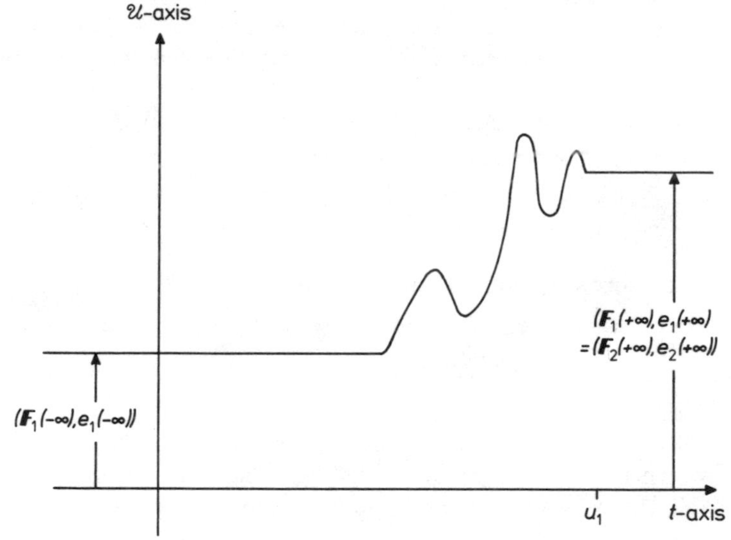

Fig. 2a: The path $(\boldsymbol{F}_1(\cdot), e_1(\cdot))$

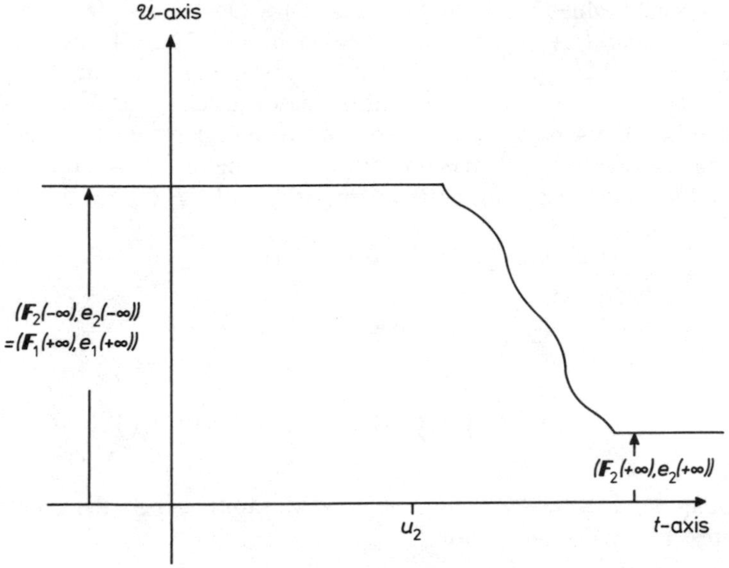

Fig. 2b: The path $(\boldsymbol{F}_2(\cdot), e_2(\cdot))$

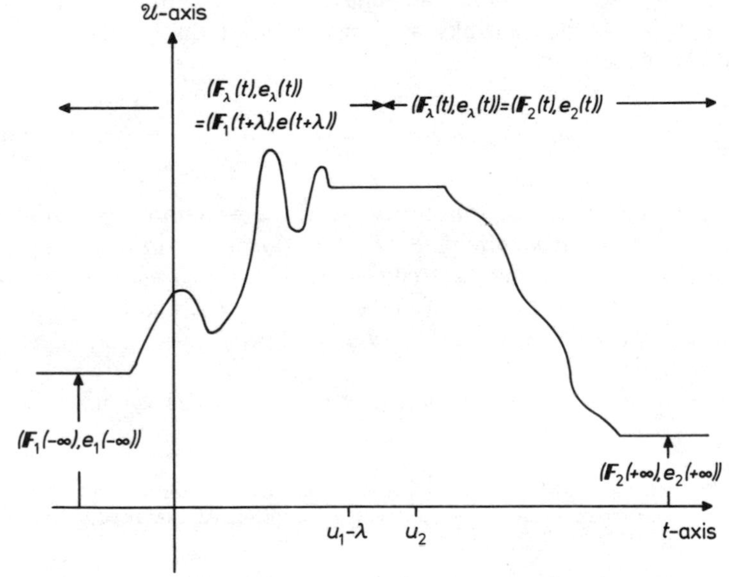

Fig. 2c: A composite path $(\boldsymbol{F}_\lambda(\cdot), e_\lambda(\cdot))$ obtained by composing the path $(\boldsymbol{F}_2(\cdot), e_2(\cdot))$ (see Fig. 2b) with the path $(\boldsymbol{F}_1(\cdot), e_1(\cdot))$ (see Fig. 2a)

and its final value is $(F_2(+\infty), e_2(+\infty))$. The paths $(F_1(\cdot), e_1(\cdot))$, $(F_2(\cdot), e_2(\cdot))$ and $(F_\lambda(\cdot), e_\lambda(\cdot))$ are indicated in Figs. 2a, 2b and 2c. It will be seen that the path $(F_\lambda(\cdot), e_\lambda(\cdot))$ is constant throughout the time interval $u_1-\lambda \leqslant t \leqslant u_2$. This interval increases in length as λ increases, that is to say the separation between the two non-constant parts of the path, one before $t=u_1-\lambda$ and one after $t=u_2$, increases as λ increases.

The Clausius integral along the composite path $(F_\lambda(\cdot), e_\lambda(\cdot))$ is

$$\begin{aligned}\mathop{\mathscr{C}}_{-\infty}^{+\infty}(F_\lambda(\cdot), e_\lambda(\cdot)) &= \mathop{\mathscr{C}}_{-\infty}^{u_1-\lambda}(F_\lambda(\cdot), e_\lambda(\cdot)) + \mathop{\mathscr{C}}_{u_2}^{+\infty}(F_\lambda(\cdot), e_\lambda(\cdot)) \\ &= \mathop{\mathscr{C}}_{-\infty}^{+\infty}(F_1(\cdot), e_1(\cdot)) \\ &\quad + \int_{u_2}^{+\infty} \mathop{\Sigma}_{s=-\infty}^{t}(F_\lambda(s)), e_\lambda(s))\cdot(\dot{F}_2(t), \dot{e}_2(t))\,dt\,.\end{aligned}$$

If the material does have a fading memory we should expect the generalised stress to satisfy the relation

$$\mathop{\Sigma}_{s=-\infty}^{t}(F_\lambda(s), e_\lambda(s)) \to \mathop{\Sigma}_{s=-\infty}^{t}(F_2(s), e_2(s))$$

as $\lambda \to +\infty$, for each fixed t in $u_2 \leqslant t < +\infty$, that is we should expect the generalised stress to forget the behaviour of the path on the interval $-\infty < t \leqslant u_1-\lambda$ which shrinks as λ increases. If that is the case we should also expect that

$$\int_{u_2}^{+\infty} \mathop{\Sigma}_{s=-\infty}^{t}(F_\lambda(s), e_\lambda(s))\cdot(\dot{F}_2(t), \dot{e}_2(t))\,dt \to \mathop{\mathscr{C}}_{u_2}^{+\infty}(F_2(\cdot), e_2(\cdot)) = \mathop{\mathscr{C}}_{-\infty}^{+\infty}(F_2(\cdot), e_2(\cdot)),$$

as $\lambda \to +\infty$, and it is just this property which we choose to adopt as our *fading memory assumption*[1]: *if* $(F_1(\cdot), e_1(\cdot))$ *and* $(F_2(\cdot), e_2(\cdot))$ *are paths with the final value of the first coinciding with the initial value of the second, that is* $(F_1(+\infty), e_1(+\infty)) = (F_2(-\infty), e_2(-\infty))$, *and if* $(F_\lambda(\cdot), e_\lambda(\cdot))$ *is the family of composite paths defined above then*

$$\mathop{\mathscr{C}}_{-\infty}^{+\infty}(F_\lambda(\cdot), e_\lambda(\cdot)) \to \mathop{\mathscr{C}}_{-\infty}^{+\infty}(F_1(\cdot), e_1(\cdot)) + \mathop{\mathscr{C}}_{-\infty}^{+\infty}(F_2(\cdot), e_2(\cdot)) \tag{3.2.1}$$

as $\lambda \to +\infty$.

---

[1] Although the relation (3.2.1) agrees with our intuition about fading memory it is not necessary to make quite such a strong assumption. All that need be assumed ist the inequality

$$\limsup_{\lambda\to+\infty} \mathop{\mathscr{C}}_{-\infty}^{+\infty}(F_\lambda(\cdot), e_\lambda(\cdot)) \geqslant \mathop{\mathscr{C}}_{-\infty}^{+\infty}(F_1(\cdot), e_1(\cdot)) + \mathop{\mathscr{C}}_{-\infty}^{+\infty}(F_2(\cdot), e_2(\cdot))$$

which avoids assuming that the limit $\lim_{\lambda\to+\infty} \mathop{\mathscr{C}}_{-\infty}^{+\infty}(F_\lambda(\cdot), e_\lambda(\cdot))$ exists.

Of course it may well be a non-trivial mathematical task to verify that a given response functional $\sum_{-\infty}^{t}$ has fading memory in this sense.

Thermoelastic materials and materials of the rate type, including linearly viscous fluids, certainly have fading memory in this sense because for them the Clausius integral along any composite path is

$$\overset{+\infty}{\underset{-\infty}{\mathscr{C}}}(\boldsymbol{F}_\lambda(\cdot), e_\lambda(\cdot)) = \overset{+\infty}{\underset{-\infty}{\mathscr{C}}}(\boldsymbol{F}_1(\cdot), e_1(\cdot)) + \overset{+\infty}{\underset{-\infty}{\mathscr{C}}}(\boldsymbol{F}_2(\cdot), e_2(\cdot)).$$

Later on we shall consider an example of a material with a memory which fades gradually rather than abruptly.

The fading memory concept introduced here is not the same as the concept introduced by Coleman and Noll [25][1]. Materials having fading memory in that sense usually have fading memory in the present sense but the converse is by no means true and the present fading memory requirement is a weaker one.

Now that we have introduced fading memory we can begin proving certain results on the way to constructing the entropy. If $(\boldsymbol{A}, a)$ and $(\boldsymbol{B}, b)$ are any pairs in the set $\mathscr{U}$ there is an infinite number of paths connecting $(\boldsymbol{A}, a)$ to $(\boldsymbol{B}, b)$. Suppose that $(\boldsymbol{F}_1(\cdot), e_1(\cdot))$ is any path with the initial value $(\boldsymbol{A}, a)$ and the final value $(\boldsymbol{B}, b)$ and that $(\boldsymbol{F}_2(\cdot), e_2(\cdot))$ is any path with the initial value $(\boldsymbol{B}, b)$ and the final value $(\boldsymbol{A}, a)$. From these paths we can construct the family of paths $(\boldsymbol{F}_\lambda(\cdot), e_\lambda(\cdot))$ as we did above. Each of these paths is closed and the Clausius inequality

$$\overset{+\infty}{\underset{-\infty}{\mathscr{C}}}(\boldsymbol{F}_\lambda(\cdot), e_\lambda(\cdot)) \leqslant 0$$

must hold. If we let $\lambda \to +\infty$ and use the fading memory assumption we find the inequality

$$\overset{+\infty}{\underset{-\infty}{\mathscr{C}}}(\boldsymbol{F}_1(\cdot), e_1(\cdot)) + \overset{+\infty}{\underset{-\infty}{\mathscr{C}}}(\boldsymbol{F}_2(\cdot), e_2(\cdot)) \leqslant 0 \tag{3.2.2}$$

or, equivalently, the inequality

$$\overset{+\infty}{\underset{-\infty}{\mathscr{C}}}(\boldsymbol{F}_1(\cdot), e_1(\cdot)) \leqslant -\overset{+\infty}{\underset{-\infty}{\mathscr{C}}}(\boldsymbol{F}_2(\cdot), e_2(\cdot)) \tag{3.2.3}$$

---

[1] See also Coleman [8, 9], Coleman and Mizel [20, 21, 22], Coleman and Noll [26], Mizel and Wang [59], Wang [70, 71] and the treatise of Truesdell and Noll [68].

which tells us that *no matter which path is chosen to connect* $(\boldsymbol{A},a)$ *to* $(\boldsymbol{B},b)$ *the Clausius integral along the path can never exceed some finite upper bound.* Let us call the least upper bound, which will depend on $(\boldsymbol{A},a)$ and on $(\boldsymbol{B},b)$, $\Pi(\boldsymbol{A},a;\,\boldsymbol{B},b)$. This means that if $(\boldsymbol{F}(\cdot),e(\cdot))$ is any path connecting $(\boldsymbol{A},a)$ to $(\boldsymbol{B},b)$ the Clausius integral satisfies the inequality

$$\mathop{\mathscr{C}}_{-\infty}^{+\infty}(\boldsymbol{F}(\cdot),e(\cdot))\leqslant\Pi(\boldsymbol{A},a;\,\boldsymbol{B},b) \tag{3.2.4}$$

and that by choosing the path suitably the Clausius integral can be made to approach $\Pi(\boldsymbol{A},a;\,\boldsymbol{B},b)$ as closely as we wish.

The inequality (3.2.2) tells us that Π has the property

$$\Pi(\boldsymbol{A},a;\,\boldsymbol{B},b)+\Pi(\boldsymbol{B},b;\,\boldsymbol{A},a)\leqslant 0 \tag{3.2.5}$$

from which it follows on setting $(\boldsymbol{B},b)=(\boldsymbol{A},a)$ that

$$\Pi(\boldsymbol{A},a;\,\boldsymbol{A},a)\leqslant 0\,. \tag{3.2.6}$$

3.3 The Entropy in Equilibrium. Thermostatics

The construction of the entropy depends crucially on the fact that, with one additional assumption on the response functional $\mathop{\Sigma}\limits_{-\infty}^{t}$ for the generalised stress, the function $\Pi(\boldsymbol{A},a;\,\boldsymbol{B},b)$ introduced at the end of section 3.2 can be decomposed into the difference

$$\Pi(\boldsymbol{A},a;\,\boldsymbol{B},b)=\eta^*(\boldsymbol{B},b)-\eta^*(\boldsymbol{A},a) \tag{3.3.1}$$

between the values of a scalar function $\eta^*(\cdot,\cdot)$ at $(\boldsymbol{B},b)$ and at $(\boldsymbol{A},a)$. Because Π can be decomposed in this way the inequalities (3.2.5) and (3.2.6) become strict equalities. The additional assumption on the generalised stress is that, in a sense we shall now explain, it must behave suitably when a path is retarded.

Let $(\boldsymbol{F}(\cdot),e(\cdot))$ be any path, let t_0 be a time with $(\boldsymbol{F}(t),e(t))\equiv(\boldsymbol{F}(-\infty),e(-\infty))$ for every $t\leqslant t_0$, let t_1 be a time with $(\boldsymbol{F}(\cdot),e(\cdot))\equiv(\boldsymbol{F}(+\infty),e(+\infty))$ for every $t\geqslant t_1$ and let α be a parameter lying in the interval $0<\alpha<1$. The path $(\boldsymbol{F}(\cdot,\alpha),e(\cdot,\alpha))$ defined by setting

$$(\boldsymbol{F}(t,\alpha),e(t,\alpha))=(\boldsymbol{F}(t_0+\alpha(t-t_0)),e(t_0+\alpha(t-t_0)))\,, \tag{3.3.2}$$

for every time t, is a *retardation* of the original path $(\boldsymbol{F}(\cdot),e(\cdot))$; it is obtained by traversing the original path at a rate reduced to a fraction α of its original value. Whereas the original process is constant on the intervals $t\leqslant t_0$ and $t\geqslant t_1$ the retarded process is constant on the intervals $t\leqslant t_0$ and $t\geqslant t_0+(1/\alpha)(t_1-t_0)$.

The Clausius integral along the retarded path is

$$\overset{+\infty}{\underset{-\infty}{\mathscr{C}}}(\boldsymbol{F}(\cdot,\alpha),e(\cdot,\alpha))$$

$$=\overset{t_0+\frac{1}{\alpha}(t_1-t_0)}{\underset{t_0}{\mathscr{C}}}(\boldsymbol{F}(\cdot,\alpha),e(\cdot,\alpha))$$

$$=\int_{t_0}^{t_0+\frac{1}{\alpha}(t_1-t_0)}\overset{t}{\underset{s=-\infty}{\Sigma}}(\boldsymbol{F}(s,\alpha),e(s,\alpha))\cdot\alpha(\dot{\boldsymbol{F}}(t_0+\alpha(t-t_0)),\dot{e}(t_0+\alpha(t-t_0))\,dt$$

and on making the change of variable $\tau=t_0+\alpha(t-t_0)$ in the integral this expression becomes

$$\overset{+\infty}{\underset{-\infty}{\mathscr{C}}}(\boldsymbol{F}(t,\alpha),e(t,\alpha))=\int_{t_0}^{t_1}\overset{t_0+\frac{1}{\alpha}(\tau-t_0)}{\underset{s=-\infty}{\Sigma}}(\boldsymbol{F}(s,\alpha),e(s,\alpha))\cdot(\dot{\boldsymbol{F}}(\tau),\dot{e}(\tau))\,d\tau. \tag{3.3.3}$$

Let us examine the generalised stress

$$\overset{t_0+\frac{1}{\alpha}(\tau-t_0)}{\underset{s=-\infty}{\Sigma}}(\boldsymbol{F}(s,\alpha),e(s,\alpha))$$

appearing in the integral on the right-hand side of equation (3.3.3). Because the variable s lies in the interval $-\infty<s\leqslant t_0+(1/\alpha)(\tau-t_0)$, the variable $u=s-t_0-(1/\alpha)(\tau-t_0)$ lies in the interval $-\infty<u\leqslant 0$ and so

$$\overset{t_0+\frac{1}{\alpha}(\tau-t_0)}{\underset{s=-\infty}{\Sigma}}(\boldsymbol{F}(s,\alpha),e(s,\alpha))$$

$$=\overset{0}{\underset{u=-\infty}{\Sigma}}\left(\boldsymbol{F}\left(t_0+\frac{1}{\alpha}(\tau-t_0)+u,\alpha\right),\,e\left(t_0+\frac{1}{\alpha}(\tau-t_0)+u,\alpha\right)\right).$$

If we now use the definition (3.3.2) of the retarded path and the requirement (3.1.8) we deduce that the generalised stress is

$$\overset{t_0+\frac{1}{\alpha}(\tau-t_0)}{\underset{s=-\infty}{\Sigma}}(\boldsymbol{F}(s,\alpha),e(s,\alpha))=\overset{0}{\underset{u=-\infty}{\Sigma}}(\boldsymbol{F}(\tau+\alpha u),e(\tau+\alpha u)). \tag{3.3.4}$$

In the limiting case of highly retarded paths, corresponding to $\alpha\to 0$,

$$(\boldsymbol{F}(\tau+\alpha u),e(\tau+\alpha u))\to(\boldsymbol{F}(\tau),e(\tau))$$

which means that for each fixed τ in $t_0<\tau<t_1$, the path $(\boldsymbol{F}(\tau+\alpha u),\,e(\tau+\alpha u))$, in which u is the independent variable, has for its pointwise limit as $\alpha\to 0$ the constant path on which the deformation gradient is always the constant tensor $\boldsymbol{F}(\tau)$ and on which the internal

energy density is always the constant scalar $e(\tau)$. In terms of the response function $\Sigma^*(\cdot,\cdot)$ for the *generalised stress in equilibrium* which is defined at pairs (A,a) in $\mathscr{U}$ as the generalised stress on the constant path (A,a), that is

$$\Sigma^*(A,a) = \mathop{\Sigma}_{-\infty}^{0}(A,a), \tag{3.3.5}$$

the generalised stress on the constant path $(F(\tau), e(\tau))$ is $\Sigma^*(F(\tau), e(\tau))$ and it is reasonable to expect that

$$\mathop{\Sigma}_{u=-\infty}^{0}(F(\tau+\alpha u), e(\tau+\alpha u)) \to \Sigma^*(F(\tau), e(\tau))$$

as $\alpha \to 0$. If this result does hold then it suggests, in conjunction with equations (3.3.3) and (3.3.4), that on highly retarded paths the Clausius integral has the limiting value

$$\mathop{\mathscr{C}}_{-\infty}^{+\infty}(F(t,\alpha), e(t,\alpha)) \to \int_{t_0}^{t_1} \Sigma^*(F(\tau), e(\tau)) \cdot (\dot{F}(\tau), \dot{e}(\tau))\, d\tau \tag{3.3.6}$$

as $\alpha \to 0$. In the language of section 1.2 the right-hand side of equation (3.3.6) is just the line integral of the ten dimensional vector field $\Sigma^*(\cdot,\cdot)$ along the original path $(F(\cdot), e(\cdot))$, for which we have the notation $I(\Sigma^*(\cdot,\cdot), F(\cdot), e(\cdot))$. It is equation (3.3.6), expressing the good behaviour of the material under retardation, which we adopt as the third, and final, restriction on the response functional $\mathop{\Sigma}_{s=-\infty}^{t}$: *if the paths* $(F(\cdot,\alpha), e(\cdot,\alpha))$ *are retardations of the path* $(F(\cdot), e(\cdot))$ *then*

$$\mathop{\mathscr{C}}_{-\infty}^{+\infty}(F(\cdot,\alpha), e(\cdot,\alpha)) \to I(\Sigma^*(\cdot,\cdot), F(\cdot), e(\cdot)) \tag{3.3.7}$$

as $\alpha \to 0$.

Any thermoelastic material always has

$$\mathop{\Sigma}_{-\infty}^{t}(F(\cdot), e(\cdot)) \equiv \Sigma^*(F(t), e(t))$$

which means that the condition (3.3.7) certainly holds for such a material. Materials of the differential type also satisfy the condition because for them the response functional for the generalised stress has the form

$$\mathop{\Sigma}_{-\infty}^{t}(F(\cdot), e(\cdot)) = \hat{\Sigma}(F(t), F^{(1)}(t), \ldots, F^{(n)}(t), e(t)),$$

where $\hat{\Sigma}$ is a continuous function, and the response function in equilibrium is

$$\Sigma^*(A,a) = \hat{\Sigma}(A, \mathbf{0}, \ldots, \mathbf{0}, a).$$

The Clausius integral along a retarded path is

$$\overset{+\infty}{\underset{-\infty}{\mathscr{C}}}(\boldsymbol{F}(\cdot,\alpha),e(\cdot,\alpha))=\int_{t_0}^{t_1}\hat{\boldsymbol{\Sigma}}(\boldsymbol{F}(t),\alpha\boldsymbol{F}^{(1)}(t),\ldots,\alpha^n\boldsymbol{F}^{(n)}(t),e(t))\cdot(\dot{\boldsymbol{F}}(t),\dot{e}(t))\,dt$$

which has the correct limit

$$\int_{t_0}^{t_1}\hat{\boldsymbol{\Sigma}}(\boldsymbol{F}(t),\boldsymbol{0},\ldots,\boldsymbol{0},e(t))\cdot(\dot{\boldsymbol{F}}(t),\dot{e}(t))\,dt=\int_{t_0}^{t_1}\boldsymbol{\Sigma}^*(\boldsymbol{F}(t),e(t))\cdot(\dot{\boldsymbol{F}}(t),\dot{e}(t))\,dt$$

as $\alpha\to 0$.

In section 4.2 we shall give examples of materials with a genuine memory which are well-behaved under retardation.

We turn to showing that *with this extra assumption the function* $\Pi(\boldsymbol{A},a;\boldsymbol{B},b)$ *must be decomposable in the form* (3.3.1).

Let $(\boldsymbol{A},a)$ and $(\boldsymbol{B},b)$ be any pairs in the set $\mathscr{U}$. If the path $(\boldsymbol{F}(\cdot),e(\cdot))$ connects $(\boldsymbol{A},a)$ to $(\boldsymbol{B},b)$ then each of the retarded paths $(\boldsymbol{F}(\cdot,\alpha),e(\cdot,\alpha))$ connects $(\boldsymbol{A},a)$ to $(\boldsymbol{B},b)$ and the definition of $\Pi(\boldsymbol{A},a;\,\boldsymbol{B},b)$ implies that

$$\overset{+\infty}{\underset{-\infty}{\mathscr{C}}}(\boldsymbol{F}(\cdot,\alpha),e(\cdot,\alpha))\leqslant\Pi(\boldsymbol{A},a;\boldsymbol{B},b)\,. \tag{3.3.8}$$

If we let $\alpha\to 0$ and use the condition (3.3.7) we deduce the inequality

$$I(\boldsymbol{\Sigma}^*(\cdot,\cdot),\boldsymbol{F}(\cdot),e(\cdot))\leqslant\Pi(\boldsymbol{A},a;\boldsymbol{B},b) \tag{3.3.9}$$

on the line integral of $\boldsymbol{\Sigma}^*(\cdot,\cdot)$. In particular if we take $(\boldsymbol{B},b)=(\boldsymbol{A},a)$ and use the property (3.2.6) of Π we deduce that if $(\boldsymbol{F}(\cdot),e(\cdot))$ is any closed path in $\mathscr{U}$ then

$$I(\boldsymbol{\Sigma}^*(\cdot,\cdot),\boldsymbol{F}(\cdot),e(\cdot))\leqslant 0\,,$$

which, as we have seen in section 1.2, tells us straightaway that $\boldsymbol{\Sigma}^*(\cdot,\cdot)$ can be expressed as the gradient of a scalar potential $\eta^*(\cdot,\cdot)$ on $\mathscr{U}$, that is

$$\boldsymbol{\Sigma}^*(\boldsymbol{F},e)=\operatorname{grad}\eta^*(\boldsymbol{F},e)=\left(\frac{\partial\eta^*}{\partial\boldsymbol{F}},\frac{\partial\eta^*}{\partial e}\right). \tag{3.3.10}$$

Since this is so the line integral on the right-hand side of (3.3.9) can be evaluated to give the inequality

$$\eta^*(\boldsymbol{B},b)-\eta^*(\boldsymbol{A},a)\leqslant\Pi(\boldsymbol{A},a;\,\boldsymbol{B},b) \tag{3.3.11}$$

which must hold for every $(\boldsymbol{A},a)$ and every $(\boldsymbol{B},b)$ in $\mathscr{U}$. On interchanging $(\boldsymbol{A},a)$ and $(\boldsymbol{B},b)$ and rearranging the inequality we find that

$$-\Pi(\boldsymbol{B},b;\,\boldsymbol{A},a)\leqslant\eta^*(\boldsymbol{B},b)-\eta^*(\boldsymbol{A},a) \tag{3.3.12}$$

and if we now combine the inequalities (3.3.11), (3.3.12) and (3.2.5) we deduce that

$$\eta^*(\boldsymbol{B},b)-\eta^*(\boldsymbol{A},a)\leqslant \Pi(\boldsymbol{A},a;\boldsymbol{B},b) \leqslant -\Pi(\boldsymbol{B},b;\boldsymbol{A},a) \leqslant \eta^*(\boldsymbol{B},b)-\eta^*(\boldsymbol{A},a)$$

which means that the decomposition (3.3.1) does hold.

The results recorded in equations (3.3.1) and (3.3.10) are extremely important. To examine the consequences of (3.3.1) let $\boldsymbol{F}_0$ be a constant tensor and let e_0 be a constant scalar with $(\boldsymbol{F}_0, e_0)$ in $\mathscr{U}$. We call $\eta^*(\boldsymbol{F}_0, e_0)$ the *equilibrium entropy* in the constant process in which the deformation gradient is $\boldsymbol{F}_0$ and the internal energy density is e_0. Equation (3.3.1) states that if $(\boldsymbol{F}(\cdot), e(\cdot))$ is any path then *the Clausius integral along the path does not exceed the change in the equilibrium entropy between the initial and final values of the path:*

$$\mathop{\mathscr{C}}_{-\infty}^{+\infty}(\boldsymbol{F}(\cdot),e(\cdot))\leqslant \eta^*(\boldsymbol{F}(+\infty),e(+\infty))-\eta^*(\boldsymbol{F}(-\infty),e(-\infty)) \quad (3.3.13)$$

and this can be written in the more familiar form

$$\int_{-\infty}^{+\infty}\frac{h(t)}{\theta(t)}dt\leqslant \eta^*(\boldsymbol{F}(+\infty),e(+\infty))-\eta^*(\boldsymbol{F}(-\infty),e(-\infty)). \quad (3.3.14)$$

In fact equation (3.3.1) states more because we have proved that the difference $\eta^*(\boldsymbol{B},b)-\eta^*(\boldsymbol{A},a)$ is the least upper bound of the Clausius integrals along all possible paths connecting $(\boldsymbol{A},a)$ to $(\boldsymbol{B},b)$, which means that *given any positive number* ε, *no matter how small, there is a path* $(\boldsymbol{F}_\varepsilon(\cdot), e_\varepsilon(\cdot))$, *depending on* ε, *with*

$$\eta^*(\boldsymbol{B},b)-\eta^*(\boldsymbol{A},a)-\varepsilon< \mathop{\mathscr{C}}_{-\infty}^{+\infty}(\boldsymbol{F}_\varepsilon(\cdot),e_\varepsilon(\cdot))\leqslant \eta^*(\boldsymbol{B},b)-\eta^*(\boldsymbol{A},a). \quad (3.3.15)$$

Indeed we know how to choose the path $(\boldsymbol{F}_\varepsilon(\cdot), e_\varepsilon(\cdot))$ because the retardations $(\boldsymbol{F}(\cdot,\alpha), e(\cdot,\alpha))$ of any path $(\boldsymbol{F}(\cdot), e(\cdot))$ connecting $(\boldsymbol{A},a)$ to $(\boldsymbol{B},b)$ have the property that as $\alpha\to 0$

$$\mathop{\mathscr{C}}_{-\infty}^{+\infty}(\boldsymbol{F}(\cdot,\alpha),e(\cdot,\alpha)\to\eta^*(\boldsymbol{B},b)-\eta^*(\boldsymbol{A},a).$$

In other words, *by taking any path connecting* $(\boldsymbol{A},a)$ *to* $(\boldsymbol{B},b)$ *and retarding it suitably we can make Clausius integral approach the change in equilibrium entropy as closely as we wish.* In this sense the processes maximising the Clausius integral along paths connecting two given pairs $(\boldsymbol{A},a)$ and $(\boldsymbol{B},b)$ in $\mathscr{U}$ are paths performed very slowly. This result gives meaning to certain remarks about "quasi-static" processes commonly encountered in works on thermodynamics.

Again to examine the relation (3.3.10) let us introduce the equilibrium response functions

$$T^*(A,a) = \overset{0}{\underset{-\infty}{T}}(A,a), \tag{3.3.16}$$

$$S^*(A,a) = \overset{0}{\underset{-\infty}{S}}(A,a), \tag{3.3.17}$$

$$\theta^*(A,a) = \overset{0}{\underset{-\infty}{\theta}}(A,a) \tag{3.3.18}$$

for the stress, the Piola-Kirchhoff stress and the absolute temperature in the constant process in which the deformation gradient is A and the internal energy density is a. Because of equations (3.1.3) and (2.2.5) the equilibrium generalised stress is

$$\begin{aligned}\Sigma^*(F,e) &= \left(-\frac{1}{\theta^*(F,e)}S^*(F,e), \frac{1}{\theta^*(F,e)}\right)\\ &= \left(-\frac{\det F}{\rho_0\,\theta^*(F,e)}T^*(F,e)(F^T)^{-1}, \frac{1}{\theta^*(F,e)}\right)\end{aligned}$$

and it follows from equation (3.3.10) that *the equilibrium entropy determines the equilibrium stress, the equilibrium Piola-Kirchhoff stress and the equilibrium temperature through the usual formulae of thermostatics,* namely

$$T^*(F,e) = \frac{-\dfrac{\rho_0}{\det F}\dfrac{\partial\eta^*}{\partial F}(F,e)\,F^T}{\dfrac{\partial\eta^*}{\partial e}(F,e)}, \tag{3.3.19}$$

$$S^*(F,e) = \frac{-\dfrac{\partial\eta^*}{\partial F}(F,e)}{\dfrac{\partial\eta^*}{\partial e}(F,e)}, \tag{3.3.20}$$

$$\theta^*(F,e) = \frac{1}{\dfrac{\partial\eta^*}{\partial e}(F,e)}. \tag{3.3.21}$$

These formulae can be written in component form, if required, by using the fact that the tensor $\partial\eta^*/\partial F$ has components

$$\left(\frac{\partial\eta^*}{\partial F}\right)_{ij} = \frac{\partial\eta^*}{\partial F_{ij}}.$$

The requirement of invariance under superposed rigid motions shows that the equilibrium entropy $\eta^*(F,e)$ can depend on F only through the symmetric Cauchy-Green tensor $C = F^T F$, that is $\eta^*(F,e) \equiv \check{\eta}(C,e)$, and the equations (3.3.19), (3.3.20) and (3.3.21) can be rewritten in terms of the derivatives of $\breve{\eta}$ with respect to $\boldsymbol{C}$ and e.

In the particular case of the compressible linearly viscous fluid the equilibrium stress is just the hydrostatic pressure $\boldsymbol{T}^* = -p(\upsilon, e)\boldsymbol{I}$, the equilibrium entropy $\eta^*(\upsilon, e)$ depends only on the specific volume υ and on e and it follows from (3.3.19) and (3.3.21) that the pressure and temperature are determined by the familiar relations

$$p = \frac{\dfrac{\partial \eta^*}{\partial \upsilon}}{\dfrac{\partial \eta^*}{\partial e}}, \qquad \theta = \frac{1}{\dfrac{\partial \eta^*}{\partial e}}. \tag{3.3.22}$$

If we regard the equation $\theta = 1/(\partial\eta^*/\partial e)$ as being inverted to express the internal energy $e = e^*(\boldsymbol{F}, \theta)$ as a function of the deformation gradient and the absolute temperature we can regard the stress, the Piola-Kirchhoff stress and the entropy in equilibrium as functions $\boldsymbol{T}^*(\boldsymbol{F}, \theta)$, $\boldsymbol{S}^*(\boldsymbol{F}, \theta)$ and $\eta^*(\boldsymbol{F}, \theta)$ of $\boldsymbol{F}$ and θ. Strictly speaking these functions should be distinguished from the functions appearing in (3.3.19), (3.3.20) and (3.3.21) but we shall not bother to introduce separate symbols for them. In terms of the *equilibrium free energy*

$$\psi^*(\boldsymbol{F}, \theta) = e^*(\boldsymbol{F}, \theta) - \theta\,\eta^*(\boldsymbol{F}, \theta) \tag{3.3.23}$$

the thermostatic relations assume the simpler forms

$$\boldsymbol{T}^*(\boldsymbol{F}, \theta) = \frac{\rho_0}{\det \boldsymbol{F}}\,\frac{\partial \psi^*}{\partial \boldsymbol{F}}(\boldsymbol{F}, \theta)\,\boldsymbol{F}^T, \tag{3.3.24}$$

$$\boldsymbol{S}^*(\boldsymbol{F}, \theta) = \frac{\partial \psi^*}{\partial \boldsymbol{F}}(\boldsymbol{F}, \theta), \tag{3.3.25}$$

$$\eta^*(\boldsymbol{F}, \theta) = -\frac{\partial \psi^*}{\partial \theta}(\boldsymbol{F}, \theta), \tag{3.3.26}$$

and the pressure and entropy in a linearly viscous fluid are

$$p = -\frac{\partial \psi^*}{\partial \upsilon}, \qquad \eta = -\frac{\partial \psi^*}{\partial \theta}. \tag{3.3.27}$$

Yet another way to write the equations is to invert the relation $\eta = \eta^*(\boldsymbol{F}, e)$ to obtain $e = e^*(\boldsymbol{F}, \eta)$ and then the thermostatic relations can be expressed as

$$\boldsymbol{T}^*(\boldsymbol{F}, \eta) = \frac{\rho_0}{\det \boldsymbol{F}}\,\frac{\partial e^*}{\partial \boldsymbol{F}}(\boldsymbol{F}, \eta)\,\boldsymbol{F}^T, \tag{3.3.28}$$

$$\boldsymbol{S}^*(\boldsymbol{F}, \eta) = \frac{\partial e^*}{\partial \boldsymbol{F}}(\boldsymbol{F}, \eta), \tag{3.3.29}$$

$$\theta^*(\boldsymbol{F}, \eta) = \frac{\partial e^*}{\partial \eta}(\boldsymbol{F}, \eta). \tag{3.3.30}$$

For the linearly viscous fluid these relations reduce to the equations

$$p = -\frac{\partial e^*}{\partial v}, \qquad \theta = \frac{\partial e^*}{\partial \eta}$$

for the pressure and the temperature.

3.4 The Entropy away from Equilibrium. The Clausius-Planck Inequality

So far the entropy has been defined only in equilibrium, that is in processes in which the deformation gradient and the internal energy density are constant. It is possible to define the entropy in more general situations; we shall define it, without making any fresh assumptions about the material, for processes which start from equilibrium, perhaps in the remote past, and in which the deformation gradient and the internal energy are subsequently allowed to vary with time in quite arbitrary ways.

The key to the construction of the entropy is to construct first a scalar functional $\overset{t}{\underset{-\infty}{H}}(\boldsymbol{F}(\cdot), e(\cdot)) = \overset{t}{\underset{s=-\infty}{H}}(\boldsymbol{F}(s), e(s))$, called the *recoverable entropy*, in the following manner. Suppose that $(\boldsymbol{F}(\cdot), e(\cdot))$ is any path and that τ is any time whatsoever. If $(\hat{\boldsymbol{F}}(\cdot), \hat{e}(\cdot))$ is a path which coincides with $(\boldsymbol{F}(\cdot), e(\cdot))$ for all times t prior to τ, that is $(\hat{\boldsymbol{F}}(t), \hat{e}(t)) \equiv (\boldsymbol{F}(t), e(t))$ for all $t \leqslant \tau$, and whose final value is $(\hat{\boldsymbol{F}}(+\infty), \hat{e}(+\infty)) = (\boldsymbol{F}(\tau), e(\tau))$ then we call it a *closed connection of the path* $(\boldsymbol{F}(\cdot), e(\cdot))$ *at the time* τ. The behaviour of the closed connection after the time τ is unrestricted except by the condition that it must eventually assume the constant final value $(\boldsymbol{F}(\tau), e(\tau))$. The simplest of all the closed connections is the *constant continuation* obtained by holding the deformation gradient and the internal energy constant after τ, that is to say $(\hat{\boldsymbol{F}}(t), \hat{e}(t)) \equiv (\boldsymbol{F}(\tau), e(\tau))$ for every $t \geqslant \tau$. Fig. 3a illustrates the concept of a closed connection of a path and Fig. 3b illustrates the constant continuation of the path.

If $(\hat{\boldsymbol{F}}(\cdot), \hat{e}(\cdot))$ is a closed connection of $(\boldsymbol{F}(\cdot), e(\cdot))$ at τ its Clausius integral must, by (3.3.13), satisfy the inequality

$$\overset{+\infty}{\underset{-\infty}{\mathscr{C}}}(\hat{\boldsymbol{F}}(\cdot), \hat{e}(\cdot)) \leqslant \eta^*(\boldsymbol{F}(\tau), e(\tau)) - \eta^*(\boldsymbol{F}(-\infty), e(-\infty)).$$

Because the paths $(\hat{\boldsymbol{F}}(\cdot), \hat{e}(\cdot))$ and $(\boldsymbol{F}(\cdot), e(\cdot))$ coincide at all times prior to τ we can write

$$\overset{+\infty}{\underset{-\infty}{\mathscr{C}}}(\hat{\boldsymbol{F}}(\cdot), \hat{e}(\cdot)) = \overset{\tau}{\underset{-\infty}{\mathscr{C}}}(\hat{\boldsymbol{F}}(\cdot), \hat{e}(\cdot)) + \overset{+\infty}{\underset{\tau}{\mathscr{C}}}(\hat{\boldsymbol{F}}(\cdot), \hat{e}(\cdot)) = \overset{\tau}{\underset{-\infty}{\mathscr{C}}}(\boldsymbol{F}(\cdot), e(\cdot)) + \overset{+\infty}{\underset{\tau}{\mathscr{C}}}(\hat{\boldsymbol{F}}(\cdot), \hat{e}(\cdot))$$

and so

$$\overset{+\infty}{\underset{\tau}{\mathscr{C}}}(\hat{\boldsymbol{F}}(\cdot), \hat{e}(\cdot)) \leqslant \eta^*(\boldsymbol{F}(\tau), e(\tau)) - \eta^*(\boldsymbol{F}(-\infty), e(-\infty)) - \overset{\tau}{\underset{-\infty}{\mathscr{C}}}(\boldsymbol{F}(\cdot), e(\cdot)). \qquad (3.4.1)$$

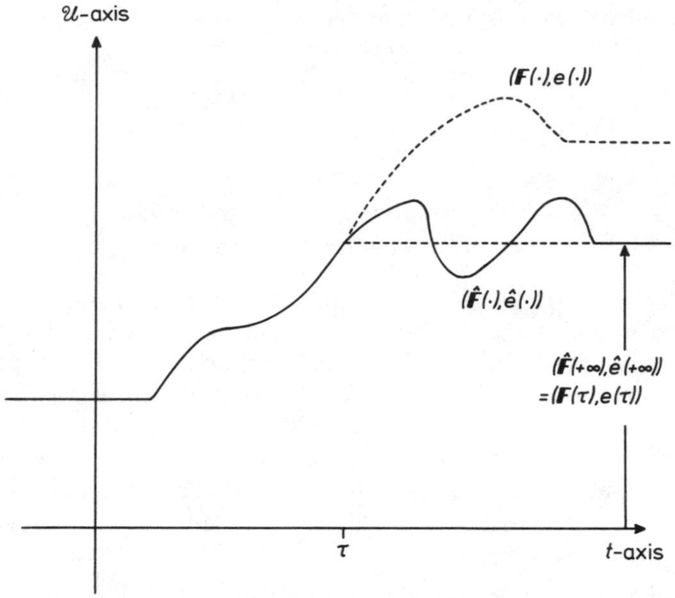

Fig. 3a: A closed connection $(\hat{\boldsymbol{F}}(\cdot), \hat{e}(\cdot))$ of the path $(\boldsymbol{F}(\cdot), e(\cdot))$ at τ

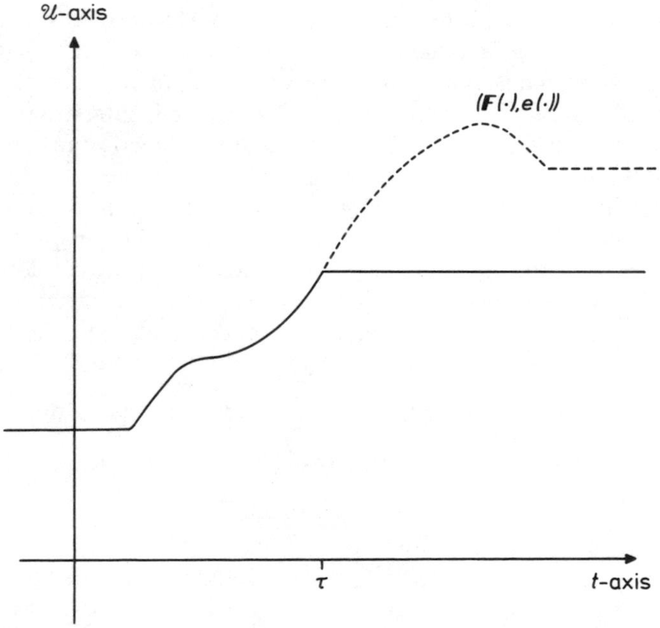

Fig. 3b: The constant continuation at τ of the path $(\boldsymbol{F}(\cdot), e(\cdot))$ (see Fig. 3a)

The value of the right-hand side of the inequality (3.4.1) is determined solely by the behaviour of the original path $(\boldsymbol{F}(\cdot), e(\cdot))$ on the time interval $-\infty < t \leqslant \tau$, whereas on the left-hand side the path $(\hat{\boldsymbol{F}}(\cdot), \hat{e}(\cdot))$ can be any closed connection of the original path at τ. Accordingly as $(\hat{\boldsymbol{F}}(\cdot), \hat{e}(\cdot))$ ranges over all closed connections at τ the Clausius integrals $\overset{+\infty}{\underset{\tau}{\mathscr{C}}}(\hat{\boldsymbol{F}}(t), \hat{e}(t))$ are bounded above by some finite number. The least upper bound will depend on the behaviour of the path $(\boldsymbol{F}(\cdot), e(\cdot))$ in the interval $-\infty < t \leqslant \tau$; it can be written as the functional

$$\overset{\tau}{\underset{-\infty}{H}}(\boldsymbol{F}(\cdot), e(\cdot)) = \overset{\tau}{\underset{s=-\infty}{H}}(\boldsymbol{F}(s), e(s))$$

and will be called the *recoverable entropy*[1] for the path $(\boldsymbol{F}(\cdot), e(\cdot))$ at time τ.

It turns out that *the recoverable entropy can never be negative:*

$$\overset{\tau}{\underset{-\infty}{H}}(\boldsymbol{F}(\cdot), e(\cdot)) \geqslant 0 \tag{3.4.2}$$

and that *no entropy is recoverable from a constant path*, that is if $(\boldsymbol{A}, a)$ is any constant pair in $\mathscr{U}$ then

$$\overset{\tau}{\underset{-\infty}{H}}(\boldsymbol{A}, a) = 0 \tag{3.4.3}$$

for any τ. To prove (3.4.2) we need only remember that the constant continuation $(\hat{\boldsymbol{F}}(\cdot), \hat{e}(\cdot))$ at time τ is a particular closed connection of the path $(\boldsymbol{F}(\cdot), e(\cdot))$ and that for this constant continuation the Clausius integral $\overset{+\infty}{\underset{\tau}{\mathscr{C}}}(\hat{\boldsymbol{F}}(\cdot), \hat{e}(\cdot))$ vanishes which means that the least upper bound $\overset{\tau}{\underset{-\infty}{H}}(\boldsymbol{F}(\cdot), e(\cdot))$ certainly cannot be negative. The proof of (3.4.3) depends on the fact that any closed connection $(\hat{\boldsymbol{F}}(\cdot), \hat{e}(\cdot))$ at time τ of a constant path $(\boldsymbol{A}, a)$ must be a closed path and, by the Clausius inequality (3.1.7),

$$\overset{+\infty}{\underset{\tau}{\mathscr{C}}}(\hat{\boldsymbol{F}}(\cdot), \hat{e}(\cdot)) = \overset{+\infty}{\underset{-\infty}{\mathscr{C}}}(\hat{\boldsymbol{F}}(\cdot), \hat{e}(\cdot)) \leqslant 0.$$

Thus the least upper bound $\overset{\tau}{\underset{-\infty}{H}}(\boldsymbol{A}, a) \leqslant 0$, whereas, according to (3.4.2), $\overset{\tau}{\underset{-\infty}{H}}(\boldsymbol{A}, a) \geqslant 0$, which proves (3.4.3).

For thermoelastic materials and for materials of the differential type the recoverable entropy always vanishes because their memory fades abruptly

---

[1] The concept of recoverable entropy defined here is closely related to the concept of maximum recoverable work which was studied for isothermal visco-elasticity, by Breuer [1], Breuer and Onat [2, 3], Day [31, 32, 34, 36, 38], Martin and Ponter [57] and, in more general contexts, by Day [29, 30].

and, consequently, if $(\hat{\boldsymbol{F}}(\cdot), \hat{e}(\cdot))$ is a closed connection of a path $(\boldsymbol{F}(\cdot), e(\cdot))$ at τ

$$\overset{+\infty}{\underset{\tau}{\mathscr{C}}}(\hat{\boldsymbol{F}}(\cdot), \hat{e}(\cdot)) = \overset{+\infty}{\underset{-\infty}{\mathscr{C}}}(\check{\boldsymbol{F}}(\cdot), \check{e}(\cdot)),$$

where $(\check{\boldsymbol{F}}(\cdot), \check{e}(\cdot))$ is the path defined by $(\check{\boldsymbol{F}}(t), \check{e}(t)) \equiv (\boldsymbol{F}(\tau), e(\tau))$ for every $t \leqslant \tau$ and $(\check{\boldsymbol{F}}(t), \check{e}(t)) \equiv (\hat{\boldsymbol{F}}(t), \hat{e}(t))$ for every $t \geqslant \tau$. But the path $(\check{\boldsymbol{F}}(\cdot), \check{e}(\cdot))$ is closed and the Clausius inequality applies to it, that is to say,

$$\overset{+\infty}{\underset{\tau}{\mathscr{C}}}(\hat{\boldsymbol{F}}(\cdot), \hat{e}(\cdot)) = \overset{+\infty}{\underset{-\infty}{\mathscr{C}}}(\check{\boldsymbol{F}}(\cdot), \check{e}(\cdot)) \leqslant 0.$$

Accordingly in a thermoelastic material or a material of the differential type

$$\overset{\tau}{\underset{-\infty}{H}}(\boldsymbol{F}(\cdot), e(\cdot)) \leqslant 0$$

and when this inequality is combined with (3.4.2) we see that the recoverable entropy $\overset{\tau}{\underset{-\infty}{H}}(\boldsymbol{F}(\cdot), e(\cdot))$ must always vanish.

In section 4.2 we shall see that there are materials in which the recoverable entropy does not always vanish. In general it is a difficult mathematical task to verify that a given response functional $\overset{t}{\underset{-\infty}{\Sigma}}$ for the generalised stress satisfies the Clausius inequality, that it has fading memory and is well-behaved under retardation and it is a difficult task too to compute the recoverable entropy $\overset{t}{\underset{-\infty}{H}}$.

Now that the equilibrium entropy and the recoverable entropy have both been defined we can define *the response functional* $\overset{t}{\underset{-\infty}{\eta}}$ *for the entropy* by

$$\overset{t}{\underset{-\infty}{\eta}}(\boldsymbol{F}(\cdot), e(\cdot)) = \overset{t}{\underset{s=-\infty}{\eta}}(\boldsymbol{F}(s), e(s)) = \eta^*(\boldsymbol{F}(t), e(t)) - \overset{t}{\underset{-\infty}{H}}(\boldsymbol{F}(\cdot), e(\cdot)). \quad (3.4.4)$$

The *entropy* at time t is $\eta(t) = \overset{t}{\underset{-\infty}{\eta}}(\boldsymbol{F}(\cdot), e(\cdot))$ and the *equilibrium entropy* is $\eta^*(\boldsymbol{F}(t), e(t))$ and so

$$\{\text{entropy}\} = \{\text{equilibrium entropy}\} - \{\text{recoverable entropy}\}.$$

Because no entropy is recoverable from a constant path $(\boldsymbol{A}, a)$ the entropy on that path is

$$\overset{0}{\underset{-\infty}{\eta}}(\boldsymbol{A}, a) = \eta^*(\boldsymbol{A}, a)$$

which means that our terminology is consistent and that *the function* $\eta^*(\cdot,\cdot)$ *introduced in section* 3.3 *is indeed the entropy in equlibrium.*

In a thermoelastic material or in a material of the differential type the recoverable entropy vanishes identically and the entropy must always coincide with the equilibrium entropy in those materials.

The fact that, in any material, the recoverable entropy can never be negative tells us that the inequality

$$\underset{-\infty}{\overset{t}{\eta}}(\boldsymbol{F}(\cdot), e(\cdot)) \leqslant \eta^*(\boldsymbol{F}(t), e(t)) \tag{3.4.5}$$

holds. One way of stating this result is to consider all possible paths with $(\boldsymbol{F}(\tau), e(\tau)) = (\boldsymbol{A}, a)$ at a given time τ. For any of these paths

$$\underset{-\infty}{\overset{\tau}{\eta}}(\boldsymbol{F}(\cdot), e(\cdot)) \leqslant \eta^*(\boldsymbol{A}, a) = \underset{-\infty}{\overset{\tau}{\eta}}(\boldsymbol{A}, a),$$

that is *among all the paths having given values of the deformation gradient and internal energy at time* τ *the constant path has the maximum entropy at that time*[1].

Perhaps the most important property of the entropy is that it satisfies the *Clausius-Planck inequality*

$$\underset{\tau_0}{\overset{\tau_1}{\mathscr{C}}}(\boldsymbol{F}(\cdot), e(\cdot)) \leqslant \underset{-\infty}{\overset{\tau_1}{\eta}}(\boldsymbol{F}(\cdot), e(\cdot)) - \underset{-\infty}{\overset{\tau_0}{\eta}}(\boldsymbol{F}(\cdot), e(\cdot)) \tag{3.4.6}$$

which can be rewritten in either of the familiar forms

$$\int_{\tau_0}^{\tau_1} \frac{h(t)}{\theta(t)}\, dt \leqslant \eta(\tau_1) - \eta(\tau_0) \tag{3.4.7}$$

or

$$0 \leqslant \eta(\tau_1) - \eta(\tau_0) + \int_{\tau_0}^{\tau_1} \frac{1}{\rho\theta}(\boldsymbol{T}\cdot\boldsymbol{D} - \rho\dot{e})\, dt. \tag{3.4.8}$$

Here τ_0 and $\tau_1 (> \tau_0)$ are any times whatsoever and $\boldsymbol{D}$ is the rate of strain tensor defined in (1.4.33).

The inequality (3.4.6) can be proved in the following way. Suppose that the path $(\boldsymbol{F}(\cdot), e(\cdot))$ and the times τ_0 and τ_1 are arbitrary, but fixed throughout the proof. Let the path $(\boldsymbol{F}_1(\cdot), e_1(\cdot))$ be any closed connection of the path $(\boldsymbol{F}(\cdot), e(\cdot))$ at the time τ_1 and let $(\boldsymbol{F}_2(\cdot), e_2(\cdot))$ be any path connecting $(\boldsymbol{F}(\tau_1), e(\tau_1))$ to $(\boldsymbol{F}(\tau_0), e(\tau_0))$ in the usual sense, namely $(\boldsymbol{F}_2(-\infty), e_2(-\infty)) = (\boldsymbol{F}(\tau_1), e(\tau_1))$ and $(\boldsymbol{F}_2(+\infty), e_2(+\infty)) = (\boldsymbol{F}(\tau_0), e(\tau_0))$. Necessarily the final value of $(\boldsymbol{F}_1(\cdot), e_1(\cdot))$ coincides with the initial value of $(\boldsymbol{F}_2(\cdot), e_2(\cdot))$ and we can form the family $(\boldsymbol{F}_\lambda(\cdot), e_\lambda(\cdot))$ of composite paths depending on the parameter λ exactly as we did in the discussion of fading memory in section 3.2. The behaviour of the path $(\boldsymbol{F}_\lambda(\cdot), e_\lambda(\cdot))$ is indicated in Figs. 4a and 4b. It is easily verified that each

[1] This result also holds in the theory of Chapter 5 based on the Clausius-Duhem inequality. It is due to Coleman [8].

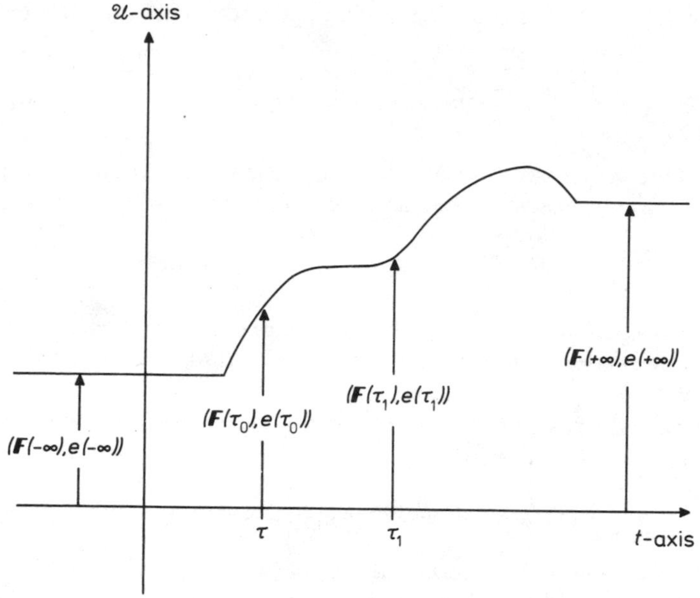

Fig. 4a: The path $(\boldsymbol{F}(\cdot), e(\cdot))$

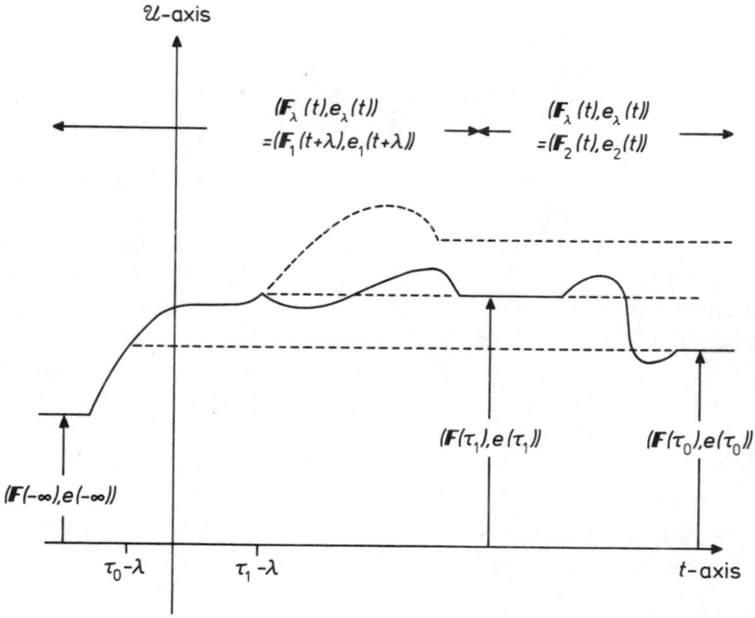

Fig. 4b: The composite paths $(\boldsymbol{F}_\lambda(\cdot), e_\lambda(\cdot))$

composite path is a closed connection of the path $(\boldsymbol{F}(t+\lambda), e(t+\lambda))$ at the time $\tau_0-\lambda$ and in consequence the definition of the recoverable entropy implies the inequality[1]

$$\overset{+\infty}{\underset{\tau_0-\lambda}{\mathscr{C}}}(\boldsymbol{F}_\lambda(\cdot), e_\lambda(\cdot)) \leqslant \overset{\tau_0-\lambda}{\underset{s=-\infty}{H}}(\boldsymbol{F}(s+\lambda), e(s+\lambda)) = \overset{\tau_0}{\underset{s=-\infty}{H}}(\boldsymbol{F}(s), e(s)).$$

Adding the term

$$\overset{\tau_0-\lambda}{\underset{-\infty}{\mathscr{C}}}(\boldsymbol{F}_\lambda(\cdot), e_\lambda(\cdot)) = \overset{\tau_0-\lambda}{\underset{t=-\infty}{\mathscr{C}}}(\boldsymbol{F}(t+\lambda), e(t+\lambda)) = \overset{+\infty}{\underset{-\infty}{\mathscr{C}}}(\boldsymbol{F}(\cdot), e(\cdot))$$

to both sides produces the inequality

$$\overset{+\infty}{\underset{-\infty}{\mathscr{C}}}(\boldsymbol{F}_\lambda(\cdot), e_\lambda(\cdot)) \leqslant \overset{\tau_0}{\underset{-\infty}{H}}(\boldsymbol{F}(\cdot), e(\cdot)) + \overset{\tau_0}{\underset{-\infty}{\mathscr{C}}}(\boldsymbol{F}(\cdot), e(\cdot))$$

and on taking the limit as $\lambda \to +\infty$ and using the fading memory assumption (3.2.1) we find the inequality

$$\overset{+\infty}{\underset{-\infty}{\mathscr{C}}}(\boldsymbol{F}_1(\cdot), e_1(\cdot)) + \overset{+\infty}{\underset{-\infty}{\mathscr{C}}}(\boldsymbol{F}_2(\cdot), e_2(\cdot)) \leqslant \overset{\tau_0}{\underset{-\infty}{H}}(\boldsymbol{F}(\cdot), e(\cdot)) + \overset{\tau_0}{\underset{-\infty}{\mathscr{C}}}(\boldsymbol{F}(\cdot), e(\cdot))$$

in which $(\boldsymbol{F}_2(\cdot), e_2(\cdot))$ can be any path connecting $(\boldsymbol{F}(\tau_1), e(\tau_1))$ to $(\boldsymbol{F}(\tau_0), e(\tau_0))$. This means that the term $\overset{+\infty}{\underset{-\infty}{\mathscr{C}}}(\boldsymbol{F}_2(\cdot), e_2(\cdot))$ can be replaced by the least upper bound as $(\boldsymbol{F}_2(\cdot), e_2(\cdot))$ varies over all paths of this kind. In other words the inequality

$$\begin{aligned} &\overset{+\infty}{\underset{-\infty}{\mathscr{C}}}(\boldsymbol{F}_1(\cdot), e_1(\cdot)) + \eta^*(\boldsymbol{F}(\tau_0), e(\tau_0)) - \eta^*(\boldsymbol{F}(\tau_1), e(\tau_1)) \\ &\qquad \leqslant \overset{\tau_0}{\underset{-\infty}{H}}(\boldsymbol{F}(\cdot), e(\cdot)) + \overset{\tau_0}{\underset{-\infty}{\mathscr{C}}}(\boldsymbol{F}(\cdot), e(\cdot)) \end{aligned} \tag{3.4.9}$$

must hold. Because the path $(\boldsymbol{F}_1(\cdot), e_1(\cdot))$ is a closed connection of the path $(\boldsymbol{F}(\cdot), e(\cdot))$ at time τ_1

$$\begin{aligned} \overset{+\infty}{\underset{-\infty}{\mathscr{C}}}(\boldsymbol{F}_1(\cdot), e_1(\cdot)) &= \left(\overset{\tau_0}{\underset{-\infty}{\mathscr{C}}} + \overset{\tau_1}{\underset{\tau_0}{\mathscr{C}}} + \overset{+\infty}{\underset{\tau_1}{\mathscr{C}}} \right)(\boldsymbol{F}_1(\cdot), e_1(\cdot)) \\ &= \overset{\tau_0}{\underset{-\infty}{\mathscr{C}}}(\boldsymbol{F}(\cdot), e(\cdot)) + \overset{\tau_1}{\underset{\tau_0}{\mathscr{C}}}(\boldsymbol{F}(\cdot), e(\cdot)) + \overset{+\infty}{\underset{\tau_1}{\mathscr{C}}}(\boldsymbol{F}_1(\cdot), e_1(\cdot)) \end{aligned}$$

and when this identity is substituted into (3.4.9) it gives the inequality

$$\begin{aligned} &\overset{\tau_1}{\underset{\tau_0}{\mathscr{C}}}(\boldsymbol{F}(\cdot), e(\cdot)) + \overset{+\infty}{\underset{\tau_1}{\mathscr{C}}}(\boldsymbol{F}_1(\cdot), e_1(\cdot)) + \eta^*(\boldsymbol{F}(\tau_0), e(\tau_0)) \\ &\qquad - \eta^*(\boldsymbol{F}(\tau_1), e(\tau_1)) \leqslant \overset{\tau_0}{\underset{-\infty}{H}}(\boldsymbol{F}(\cdot), e(\cdot)) \end{aligned}$$

in which $(\boldsymbol{F}_1(\cdot), e_1(\cdot))$ is any closed connection of the path $(\boldsymbol{F}(\cdot), e(\cdot))$ at the time τ_1. If we now vary $(\boldsymbol{F}_1(\cdot), e_1(\cdot))$ over all closed connections of

---

[1] It follows from (3.1.8) that the functional $\overset{t}{\underset{-\infty}{H}}$ is invariant under change of origin of the time-scale.

this kind and use the definition of the recoverable entropy we are left with the inequality

$$\overset{\tau_1}{\underset{\tau_0}{\mathscr{C}}}(\boldsymbol{F}(\cdot),e(\cdot))+\overset{\tau_1}{\underset{-\infty}{H}}(\boldsymbol{F}(\cdot),e(\cdot))+\eta^*(\boldsymbol{F}(\tau_0),e(\tau_0))$$
$$-\eta^*(\boldsymbol{F}(\tau_1),e(\tau_1))\leqslant\overset{\tau_0}{\underset{-\infty}{H}}(\boldsymbol{F}(\cdot),e(\cdot))$$

which can be rearranged as the inequality

$$\overset{\tau_1}{\underset{\tau_0}{\mathscr{C}}}(\boldsymbol{F}(\cdot),e(\cdot))\leqslant\left(\eta^*(\boldsymbol{F}(\tau_1),e(\tau_1))-\overset{\tau_1}{\underset{-\infty}{H}}(\boldsymbol{F}(\cdot),e(\cdot))\right)$$
$$-\left(\eta^*(\boldsymbol{F}(\tau_0),e(\tau_0))-\overset{\tau_0}{\underset{-\infty}{H}}(\boldsymbol{F}(\cdot),e(\cdot))\right)$$

and this, because of the definition (3.4.4) of the entropy, is just the Clausius-Planck inequality (3.4.6).

The Clausius-Planck inequality enables us to deduce further properties of the entropy in a straightforward way. For example suppose that $(\boldsymbol{F}(\cdot),e(\cdot))$ is any path and that it assumes its initial value at all times before t_0 and its final value at all times after t_1. Then the entropy $\eta(t)$ must have the constant value $\eta^*(\boldsymbol{F}(-\infty),e(-\infty))$ for every time $t\leqslant t_0$. Also the Clausius integral $\overset{\tau_1}{\underset{\tau_0}{\mathscr{C}}}(\boldsymbol{F}(\cdot),e(\cdot))$ vanishes if $t_1<\tau_0<\tau_1$ and the inequality implies that $\eta(\tau_1)\geqslant\eta(\tau_0)$, that is to say the entropy $\eta(t)$ is an increasing[1] function for times $t\geqslant t_1$: *if the deformation gradient and the internal energy are held fixed (and hence the total heat supply h vanishes) the entropy cannot decrease.* As we shall see in section 4.2, in materials with a genuine memory, such as linear viscoelastic materials, the entropy can actually increase in these circumstances but in thermoelastic materials or materials of the differential type the entropy merely remains constant.

Since $(\boldsymbol{F}(t),e(t))\equiv(\boldsymbol{F}(+\infty),e(+\infty))$ for all times $t\geqslant t_1$ the inequality (3.4.6) implies that $\eta(t)\leqslant\eta^*(\boldsymbol{F}(+\infty),e(+\infty))$ for every $t\geqslant t_1$ and consequently the limit $\lim_{t\to+\infty}\eta(t)$ exists and cannot exceed the equilibrium entropy $\eta^*(\boldsymbol{F}(+\infty),e(+\infty))$ corresponding to the final value of the path. It seems entirely reasonable to expect, as a consequence of the fading memory of the material, that

$$\lim_{t\to+\infty}\eta(t)=\eta^*(\boldsymbol{F}(+\infty),e(+\infty)) \tag{3.4.10}$$

although the fading memory assumption made in section 3.2 does not appear to be strong enough for us to deduce this.

Between the times t_0 and t_1 the entropy may increase or decrease depending on the behaviour of the path. It is frequently asserted that

[1] The adjective "increasing" is used in the broad sense to mean "non-decreasing".

thermodynamics reflects the idea that in some sense there is a "trend in time" in physical processes. The result that the entropy cannot decrease when the deformation gradient and the internal energy are held fixed can be regarded as giving some precision to this idea. Another, and more general, way of making it precise is to associate with each path the *trend*

$$\xi(t) = \underset{-\infty}{\overset{t}{\eta}}(\boldsymbol{F}(\cdot), e(\cdot)) - \eta^*(\boldsymbol{F}(-\infty), e(-\infty)) - \underset{-\infty}{\overset{t}{\mathscr{C}}}(\boldsymbol{F}(\cdot), e(\cdot)) \tag{3.4.11}$$

$$= \eta(t) - \eta(-\infty) - \int_{-\infty}^{t} \frac{h(s)}{\theta(s)}\, ds\,. \tag{3.4.12}$$

Since

$$\underset{\tau_0}{\overset{\tau_1}{\mathscr{C}}}(\boldsymbol{F}(\cdot), e(\cdot)) = \underset{-\infty}{\overset{\tau_1}{\mathscr{C}}}(\boldsymbol{F}(\cdot), e(\cdot)) - \underset{-\infty}{\overset{\tau_0}{\mathscr{C}}}(\boldsymbol{F}(\cdot), e(\cdot))$$

the Clausius-Planck inequality (3.4.6) tells us that $\xi(\tau_0) \leqslant \xi(\tau_1)$ whenever $\tau_0 < \tau_1$. It also tells us that

$$\underset{-\infty}{\overset{t}{\mathscr{C}}}(\boldsymbol{F}(\cdot), e(\cdot)) \leqslant \underset{-\infty}{\overset{t}{\eta}}(\boldsymbol{F}(\cdot), e(\cdot)) - \eta^*(\boldsymbol{F}(-\infty), e(-\infty))$$

and so $\xi(t) \geqslant 0$ for every time t. Thus *on each path the trend* $\xi(\cdot)$ *is an increasing function which is never negative.* For times $t \leqslant t_0$ the trend vanishes and if the entropy has the property (3.4.10)

$$\xi(t) \to \eta^*(+\infty) - \eta^*(-\infty) - \int_{-\infty}^{+\infty} \frac{h(s)}{\theta(s)}\, ds \tag{3.4.13}$$

as $t \to +\infty$. *On any constant path the trend vanishes identically.*

If it happens that the trend and the entropy are differentiable functions[1] of the time then differentiating both sides of the definition (3.4.10) produces the inequality

$$\dot{\eta} - \frac{h}{\theta} = \dot{\xi} \geqslant 0 \tag{3.4.14}$$

which is just the local form of the inequality (3.4.7). The local form of (3.4.8) is of course

$$\rho \dot{e} \leqslant \rho \theta \dot{\eta} + \boldsymbol{T} \cdot \boldsymbol{D}\,. \tag{3.4.15}$$

---

[1] Since the trend $\xi(\cdot)$ is increasing it must be differentiable almost everywhere. The definition (3.4.12) then implies that the entropy $\eta(\cdot)$ is differentiable almost everywhere.

CHAPTER 4

Applications

For most simple materials with memory it is a difficult task to verify that the thermodynamic inequality and the assumptions about fading memory and about behaviour under retardation are satisfied and even when these assumptions can be verified it appears to be an even more difficult task to find explicit expressions for the entropy. It happens though that these tasks can be performed without undue labour for thermoelastic materials, for materials of the differential type and for certain viscoelastic materials. This Chapter is devoted to verifying the assumptions and to computing the entropy for these materials and in this way it provides examples and illustrations of the general arguments given in Chapter 3. In essence we shall still be examining the restrictions which the Clausius inequality (3.1.7) imposes on the response functionals $\overset{t}{\underset{-\infty}{\boldsymbol{T}}}$ and $\overset{t}{\underset{-\infty}{\theta}}$ for the stress and the temperature. We have seen already, in the discussion of heat conduction inequalities in section 2.3, how the thermodynamic inequality (2.2.3) restricts the response functional $\overset{t}{\underset{-\infty}{\boldsymbol{q}}}$ for the heat flux vector and we shall say nothing more about restrictions on the heat flux.

4.1 Thermoelasticity and Materials of Differential Type

Thermoelastic materials can be discussed very briefly. In a trivial way a thermoelastic material has fading memory and it is well-behaved under retardation in the sense in which these ideas were used in sections 3.2 and 3.3. If the material meets the thermodynamic requirement (2.2.3) no entropy can be recovered from any path and this means, as we should expect, that the entropy always coincides with the entropy in equilibrium:

$$\overset{t}{\underset{-\infty}{\eta}}(\boldsymbol{F}(\cdot), e(\cdot)) \equiv \eta^*(\boldsymbol{F}(t), e(t)) . \tag{4.1.1}$$

Since the stress and the temperature coincide with their values in equilibrium they can be computed from the entropy using the thermostatic formulae found in section 3.3, namely

$$T^*(F,e) = -\frac{\rho_0}{\det F}\frac{\partial\eta^*}{\partial F}(F,e)F^T \quad \frac{\partial\eta^*}{\partial e}(F,e) \tag{4.1.2}$$

and

$$\theta^*(F,e) = \frac{1}{\dfrac{\partial\eta^*}{\partial e}(F,e)}. \tag{4.1.3}$$

For thermoelastic materials each of the three equivalent forms (3.4.6), (3.4.7) and (3.4.8) of the Clausius-Planck inequality becomes an equality and the trend $\xi(\cdot)$ always vanishes identically.

Next let us discuss materials of the differential type and of complexity n and whose constitutive equations are (1.4.25), (1.4.26) and (1.4.27). The generalised stress is determined by a response function

$$\hat{\Sigma}(F, F^{(1)}, \ldots, F^{(n)}, e)$$

and in equilibrium by the function

$$\Sigma^*(F,e) = \hat{\Sigma}(F, 0, \ldots, 0, e). \tag{4.1.4}$$

We know already that this material has fading memory, that it is well-behaved under retardation, that the entropy coincides with the entropy $\eta^*(\cdot,\cdot)$ in equilibrium (that is (4.1.1) holds) and that the entropy determines the generalised stress in equilibrium through the relation

$$\Sigma^*(F,e) = \left(\frac{\partial\eta^*}{\partial F}(F,e), \frac{\partial\eta^*}{\partial e}(F,e)\right). \tag{4.1.5}$$

This last formula implies the expressions (4.1.2) and (4.1.3) for the stress and for the temperature in equilibrium.

Because of equations (4.1.1) and (4.1.5) the change in entropy between any two times τ_0 and τ_1 on any path is

$$\eta(\tau_1) - \eta(\tau_0) = \int_{\tau_0}^{\tau_1} \Sigma^*(F(t), e(t)) \cdot (F^{(1)}(t), e^{(1)}(t))\, dt$$

and on substituting this expression into the Clausius-Planck inequality (3.4.6) we find that the inequality

$$\int_{\tau_0}^{\tau_1} \hat{\Sigma}(F(t), F^{(1)}(t), \ldots, F^{(n)}(t), e(t)) \cdot (F^{(1)}(t), e^{(1)}(t))\, dt$$
$$\leqslant \int_{\tau_0}^{\tau_1} \Sigma^*(F(t), e(t)) \cdot (F^{(1)}(t), e^{(1)}(t))\, dt$$

must hold for any times τ_0 and τ_1 $(>\tau_0)$ whatsoever. This can be true only if the difference

$$\boldsymbol{\Sigma}_d(\boldsymbol{F},\boldsymbol{F}^{(1)},\ldots,\boldsymbol{F}^{(n)}\,e)=\hat{\boldsymbol{\Sigma}}(\boldsymbol{F},\boldsymbol{F}^{(1)},\ldots,\boldsymbol{F}^{(n)},e)-\boldsymbol{\Sigma}^*(\boldsymbol{F},e)\,, \tag{4.1.6}$$

which is the "dynamic" or "extra" generalised stress over and above the equilibrium generalised stress and arising from the non-vanishing of the rates $\boldsymbol{F}^{(1)},\boldsymbol{F}^{(2)},\ldots,\boldsymbol{F}^{(n)}$, satisfies the inequality

$$\boldsymbol{\Sigma}_d(\boldsymbol{F},\boldsymbol{F}^{(1)},\ldots,\boldsymbol{F}^{(n)},e)\cdot(\boldsymbol{F}^{(1)},e^{(1)})\leqslant 0 \tag{4.1.7}$$

on any path. If $\boldsymbol{M}_0,\boldsymbol{M}_1,\ldots,\boldsymbol{M}_n$ are any tensors and α and β are any scalars with the pair $(\boldsymbol{M}_0,\alpha)$ lying in the set $\mathscr{U}$ there is a path[1] $(\boldsymbol{F}(\cdot).e(\cdot))$ for which

$$\boldsymbol{F}^{(k)}(t_0)=\boldsymbol{M}_k \qquad (k=0,1,\ldots,n), \qquad e(t_0)=\alpha, \qquad e^{(1)}(t_0)=\beta \tag{4.1.8}$$

at any prescribed time t_0 and so

$$\boldsymbol{\Sigma}_d(\boldsymbol{M}_0,\boldsymbol{M}_1,\ldots,\boldsymbol{M}_n,\alpha)\cdot(\boldsymbol{M}_1,\beta)\leqslant 0\,. \tag{4.1.9}$$

But the generalised stress was defined in (3.1.3) to be

$$\boldsymbol{\Sigma}=\left(-\frac{1}{\theta}\boldsymbol{S},\frac{1}{\theta}\right)$$

which means that the extra generalised stress is

$$\boldsymbol{\Sigma}_d=\left(\frac{\boldsymbol{S}^*}{\theta^*}-\frac{\boldsymbol{S}}{\theta},\frac{1}{\theta}-\frac{1}{\theta^*}\right)$$

and consequently the inequality (4.1.9) states that

$$\begin{aligned}&\left(\frac{\boldsymbol{S}^*(\boldsymbol{M}_0,\alpha)}{\theta^*(\boldsymbol{M}_0,\alpha)}-\frac{\boldsymbol{S}(\boldsymbol{M}_0,\boldsymbol{M}_1,\ldots,\boldsymbol{M}_n,\alpha)}{\theta(\boldsymbol{M}_0,\boldsymbol{M}_1,\ldots,\boldsymbol{M}_n,\alpha)}\right)\cdot\boldsymbol{M}_1\\&\qquad+\left(\frac{1}{\theta(\boldsymbol{M}_0,\boldsymbol{M}_1,\ldots,\boldsymbol{M}_n,\alpha)}-\frac{1}{\theta^*(\boldsymbol{M}_0,\alpha)}\right)\beta\leqslant 0\,.\end{aligned} \tag{4.1.10}$$

---

[1] One way of constructing a path with the required properties is to introduce the functions

$$f_k(t)=\frac{t^k}{k!(1+t^{2n})} \qquad (k=1,2,\ldots,n)$$

which are uniformly bounded by a constant depending only on the complexity n of the material and for which $f_k^{(i)}(0)=0$ if $i\leqslant k-1$ or if $k+1\leqslant i\leqslant n$ and $f_k^{(k)}(0)=1$. If we set

$$\boldsymbol{F}(t)=\boldsymbol{M}_0+\sum_{k=1}^{n}\varepsilon^k f_k\left(\frac{1}{\varepsilon}(t-t_0)\right)\boldsymbol{M}_k$$

$$e(t)=\alpha+\varepsilon f_1\left(\frac{1}{\varepsilon}(t-t_0)\right)\beta$$

the conditions (4.1.8) are satisfied and if the constant ε is sufficiently small the vector $(\boldsymbol{F}(t),e(t))$ always lies in the open set $\mathscr{U}$ whatever the value of t.

This inequality can hold for all choices of the scalar β only if

$$\theta(\boldsymbol{M}_0, \boldsymbol{M}_1, \ldots, \boldsymbol{M}_n, \alpha) \equiv \theta^*(\boldsymbol{M}_0, \alpha). \tag{4.1.11}$$

In other words *the temperature* $\theta(\boldsymbol{F}, \boldsymbol{F}^{(1)}, \ldots, \boldsymbol{F}^{(n)}, e)$ *in a material of the differential type must be independent of the rates* $\boldsymbol{F}^{(1)}, \boldsymbol{F}^{(2)}, \ldots, \boldsymbol{F}^{(n)}$ *and it coincides with the temperature* $\theta^*(\boldsymbol{F}, e)$ *in equilibrium.*

On substituting from (4.1.11) back into the inequality (4.1.10) it simplifies to

$$(\boldsymbol{S}(\boldsymbol{M}_0, \boldsymbol{M}_1, \ldots, \boldsymbol{M}_n, \alpha) - \boldsymbol{S}^*(\boldsymbol{M}_0, \alpha)) \cdot \boldsymbol{M}_1 \geqslant 0$$

or, equivalently, to the inequality

$$(\boldsymbol{S}(\boldsymbol{F}, \boldsymbol{F}^{(1)}, \ldots, \boldsymbol{F}^{(n)}, e) - \boldsymbol{S}^*(\boldsymbol{F}, e)) \cdot \boldsymbol{F}^{(1)} \geqslant 0 \tag{4.1.12}$$

for the Piola-Kirchhoff stress. In terms of the symmetric stress tensor $\boldsymbol{T}$ and the rate of strain tensor $\boldsymbol{D}$ this inequality becomes

$$(\boldsymbol{T}_0(\boldsymbol{F}, \boldsymbol{F}^{(1)}, \ldots, \boldsymbol{F}^{(n)}, e) - \boldsymbol{T}^*(\boldsymbol{F}, e)) \cdot \boldsymbol{D} \geqslant 0 \tag{4.1.13}$$

and so *the rate at which the extra stress does mechanical work is never negative.*

In the very special case of the compressible linearly viscous fluid the stress is

$$\boldsymbol{T} = -p(v, e)\boldsymbol{I} + \lambda(v, e)(\operatorname{trace} \boldsymbol{D})\boldsymbol{I} + 2\mu(v, e)\boldsymbol{D},$$

the stress in equilibrium is the hydrostatic pressure

$$\boldsymbol{T}^* = -p(v, e)\boldsymbol{I}$$

and the inequality (4.1.13) tells us that the viscosities satisfy the inequality

$$\lambda(\operatorname{trace} \boldsymbol{D})^2 + 2\mu \operatorname{trace} \boldsymbol{D}^2 \geqslant 0 \tag{4.1.14}$$

in which $\boldsymbol{D}$ can be any symmetric tensor[1]. On choosing

$$\boldsymbol{D} = \begin{pmatrix} 1 & 0 & 0 \\ 0 & -1 & 0 \\ 0 & 0 & 0 \end{pmatrix}$$

---

[1] If $\boldsymbol{D}$ is any constant symmetric tensor and if v and e are any positive constant scalars the path defined by

$$\boldsymbol{F}(t) = (\rho_0 v)^{\frac{1}{3}} \left(\boldsymbol{I} + \varepsilon f_1 \left(\frac{1}{\varepsilon} t \right) \boldsymbol{D} \right), \qquad e(t) \equiv e$$

is one on which at $t=0$ the rate of strain tensor is $\boldsymbol{D}$, the specific volume is v and the internal energy density is e.

we deduce, as we should expect, that *the shear viscosity cannot be negative*, that is

$$\mu \geqslant 0 \tag{4.1.15}$$

and if we take $\boldsymbol{D}$ to be the identity tensor $\boldsymbol{I}$ we find that *the bulk viscosity cannot be negative:*

$$\lambda + \tfrac{2}{3}\mu \geqslant 0\,. \tag{4.1.16}$$

Since

$$\begin{aligned}&\lambda(\operatorname{trace}\boldsymbol{D})^2 + 2\mu \operatorname{trace}\boldsymbol{D}^2\\ &= (\lambda + \tfrac{2}{3}\mu)(\operatorname{trace}\boldsymbol{D})^2 + 2\mu(\operatorname{trace}\boldsymbol{D}^2 - \tfrac{1}{3}(\operatorname{trace}\boldsymbol{D})^2)\end{aligned}$$

and since[1]

$$(\operatorname{trace}\boldsymbol{D})^2 \leqslant 3\operatorname{trace}\boldsymbol{D}^2$$

the restrictions (4.1.15) and (4.1.16) on the shear and bulk viscosities are necessary and sufficient for the inequality (4.1.14) to hold.

In a material of the differential type the trend is

$$\begin{aligned}\xi(t) = \eta(t) - \eta(-\infty) - \int_{-\infty}^{t} \frac{h(s)}{\theta(s)}\,ds &= \int_{-\infty}^{t} \boldsymbol{\Sigma}^*(\boldsymbol{F}(s), e(s))\cdot(\boldsymbol{F}^{(1)}(s), e^{(1)}(s))\,ds\\ &\quad - \int_{-\infty}^{t} \hat{\boldsymbol{\Sigma}}(\boldsymbol{F}(s), \boldsymbol{F}^{(1)}(s), \ldots, \boldsymbol{F}^{(n)}(s), e(s))\cdot(\boldsymbol{F}^{(1)}(s), e^{(1)}(s))\,ds\\ &= -\int_{-\infty}^{t} \boldsymbol{\Sigma}_d(s)\cdot(\boldsymbol{F}^{(1)}(s), e^{(1)}(s))\,ds\end{aligned}$$

which, because of (4.1.11), reduces to

$$\xi(t) = \int_{-\infty}^{t} \frac{1}{\rho(s)\theta(s)}(\boldsymbol{T}(s) - \boldsymbol{T}^*(s))\cdot\boldsymbol{D}(s)\,ds \tag{4.1.17}$$

and in the particular case of the linearly viscous fluid to

$$\xi(t) = \int_{-\infty}^{t} \frac{\upsilon(s)}{\theta(s)}\{\lambda(\operatorname{trace}\boldsymbol{D}(s))^2 + 2\mu\operatorname{trace}\boldsymbol{D}(s)^2\}\,ds\,. \tag{4.1.18}$$

The fact expressed by the inequalities (4.1.13) and (4.1.14), namely the rate at which the extra stress does work cannot be negative, clearly implies that the trend increases with time for these materials.

---

[1] This inequality is just the Schwarz inequality

$$(\boldsymbol{D}\cdot\boldsymbol{I})^2 \leqslant (\boldsymbol{I}\cdot\boldsymbol{I})(\boldsymbol{D}\cdot\boldsymbol{D})$$

corresponding to the particular scalar product $\boldsymbol{A}\cdot\boldsymbol{B} = \operatorname{trace}\boldsymbol{A}\boldsymbol{B}^T$.

4.2 A Class of Viscoelastic Materials

It happens that for certain viscoelastic materials the assumptions about fading memory and behaviour under retardation can be verified and the entropy can be computed. The constitutive equations used here have been chosen because of their mathematical convenience; it is not claimed that they necessarily apply to some material occurring in nature, although they may do. Our aim is to illustrate the qualitative behaviour of materials with a genuine memory and to show how they differ from, let us say, materials of the differential type.

We shall assume that all the displacement gradients measured from some reference configuration of the material are suitably small. If $\boldsymbol{u}(\boldsymbol{X},t)$ is the displacement suffered by a particle $\boldsymbol{X}$ of the material at the time t the deformation gradient $\boldsymbol{F}$ is related to the displacement gradient by

$$\boldsymbol{F}=\boldsymbol{I}+\mathrm{GRAD}\,\boldsymbol{u} \tag{4.2.1}$$

which means that the Cauchy-Green tensor is

$$\boldsymbol{C}=\boldsymbol{F}^T\boldsymbol{F}=\boldsymbol{I}+\mathrm{GRAD}\,\boldsymbol{u}+(\mathrm{GRAD}\,\boldsymbol{u})^T+(\mathrm{GRAD}\,\boldsymbol{u})^T(\mathrm{GRAD}\,\boldsymbol{u})$$

and that the infinitesimal strain tensor

$$\boldsymbol{E}=\tfrac{1}{2}[\mathrm{GRAD}\,\boldsymbol{u}+(\mathrm{GRAD}\,\boldsymbol{u})^T] \tag{4.2.2}$$

whose components are

$$E_{ij}=\tfrac{1}{2}\left(\frac{\partial u_i}{\partial X_j}+\frac{\partial u_j}{\partial X_i}\right)$$

is a good approximation to the tensor $\tfrac{1}{2}(\boldsymbol{C}-\boldsymbol{I})$ when the displacement gradients are small in an appropriate sense. The tensor $\boldsymbol{E}$ is symmetric; when we speak of *strain space* we shall mean the collection of all symmetric tensors. In the approximation of small displacement gradients the mass density ρ can be replaced by the density ρ_0 in the reference configuration and the rate of strain tensor $\boldsymbol{D}$ by $\dot{\boldsymbol{E}}$ so that equation (1.3.15) becomes

$$\dot{e}=\frac{1}{\rho}\,\boldsymbol{T}\cdot\dot{\boldsymbol{E}}+h\,. \tag{4.2.3}$$

The materials we consider are those in which the stress, the temperature and the heat flux are given by the equations

$$\boldsymbol{T}=\rho_0\,\theta\,\boldsymbol{\Delta}, \tag{4.2.4}$$

$$\theta=\theta(e), \tag{4.2.5}$$

$$\boldsymbol{q}=-\kappa\,\boldsymbol{g} \tag{4.2.6}$$

where the thermal conductivity κ is a positive constant and $\Delta(\cdot)$ is the linear hereditary functional

$$\Delta(t)=\Delta(t,\boldsymbol{E}(\cdot))=\mathscr{G}(0)\boldsymbol{E}(t)+\int_{-\infty}^{t}\dot{\mathscr{G}}(t-s)\boldsymbol{E}(s)\,ds\,. \tag{4.2.7}$$

Here $\mathscr{G}(\cdot)$ is the relaxation function. Its values are fourth order tensors whose cartesian components are $\mathscr{G}_{ijkl}(\cdot)$, which means that (4.2.7) can be written in component form as

$$\Delta_{ij}(t,\boldsymbol{E}(\cdot))=\sum_{k,l=1,2,3}\left\{\mathscr{G}_{ijkl}(0)\,E_{kl}(t)+\int_{-\infty}^{t}\dot{\mathscr{G}}_{ijkl}(t-s)\,E_{kl}(s)\,ds\right\}. \tag{4.2.8}$$

Because the stress tensor $\boldsymbol{T}$ and the infinitesimal strain tensor $\boldsymbol{E}$ are both symmetric we can take it that the relaxation function has the symmetries

$$\mathscr{G}_{ijkl}(\cdot)=\mathscr{G}_{jikl}(\cdot), \qquad \mathscr{G}_{ijkl}(\cdot)=\mathscr{G}_{ijlk}(\cdot) \tag{4.2.9}$$

and this means that at most 36 of the 81 functions $\mathscr{G}_{ijkl}(\cdot)$ are independent of each other. In fact we shall, for simplicity, confine our attention to those relaxation functions which are symmetric[1] in the sense that in addition to the restrictions (4.2.9) they satisfy

$$\mathscr{G}_{ijkl}(\cdot)=\mathscr{G}_{klij}(\cdot)\,. \tag{4.2.10}$$

In view of these extra conditions only 21 of the functions $\mathscr{G}_{ijkl}(\cdot)$ can be independent and if $\boldsymbol{A}$ and $\boldsymbol{B}$ are any symmetric tensors

$$\boldsymbol{A}\cdot\mathscr{G}(\cdot)\boldsymbol{B}=\boldsymbol{B}\cdot\mathscr{G}(\cdot)\boldsymbol{A}\,, \tag{4.2.11}$$

that is to say

$$\sum_{i,j,k,l}A_{ij}B_{kl}\mathscr{G}_{ijkl}(\cdot)=\sum_{i,j,k,l}B_{ij}A_{kl}\mathscr{G}_{ijkl}(\cdot)\,.$$

By the *norm* of the relaxation function at time t is meant the finite number $\|\mathscr{G}(t)\|$, which is defined to be the least upper bound of the numbers

$$\frac{|\mathscr{G}(t)\boldsymbol{L}|}{|\boldsymbol{L}|}$$

as $\boldsymbol{L}$ ranges over all non-zero symmetric tensors.

The material defined by equations (4.2.4), (4.2.5), (4.2.6) and (4.2.7) is usually anisotropic. For it to be isotropic the relaxation function must assume the very special form

$$\mathscr{G}_{ijkl}(\cdot)=\lambda(\cdot)\,\delta_{ij}\delta_{kl}+\mu(\cdot)(\delta_{ik}\delta_{jl}+\delta_{il}\delta_{kj}) \tag{4.2.12}$$

---

[1] Cf. section 6.2.

where δ_{ij} is the Kronecker delta with $\delta_{ij}=1$ if $i=j$ and $\delta_{ij}=0$ if $i \neq j$, $\mu(\cdot)$ is the shear relaxation function and $\lambda(\cdot)+\frac{2}{3}\mu(\cdot)$ the bulk relaxation function. In this case the relation (4.2.7) reduces to the form

$$\begin{aligned} \boldsymbol{\Delta}(t,\boldsymbol{E}(\cdot)) &= \lambda(0)(\operatorname{trace}\boldsymbol{E}(t))\boldsymbol{I}+2\mu(0)\boldsymbol{E}(t) \\ &+ \int_{-\infty}^{t} \{\dot{\lambda}(t-s)(\operatorname{trace}\boldsymbol{E}(s))\boldsymbol{I}+2\dot{\mu}(t-s)\boldsymbol{E}(s)\}\,ds\,. \end{aligned} \tag{4.2.13}$$

To ensure the validity of the arguments which follow we shall require the relaxation function $\mathscr{G}(\cdot)$ to meet certain requirements. One requirement is that the limit $\mathscr{G}(+\infty)=\lim_{s\to+\infty}\mathscr{G}(s)$ shall exists; if it does the response function for the stress in equilibrium is

$$\boldsymbol{T}^* = \rho_0\,\theta\,\boldsymbol{\Delta}^* \tag{4.2.14}$$

where

$$\boldsymbol{\Delta}^*(\boldsymbol{E}) = \mathscr{G}(+\infty)\boldsymbol{E}. \tag{4.2.15}$$

Next we turn to verifying that the assumptions made in Chapter 3 do hold. The integrand appearing in the thermodynamic inequality (2.2.3) is[1]

$$-\frac{1}{\rho_0}\operatorname{div}\left(\frac{1}{\theta}\boldsymbol{q}\right)+\frac{1}{\theta}r=\frac{1}{\theta}h+\frac{1}{\rho_0\theta^2}\boldsymbol{q}\cdot\boldsymbol{g}$$

and, because of the constitutive relations (4.2.4), (4.2.5) and (4.2.6) and equation (4.2.3) and the fact that the thermal conductivity κ is positive, we have

$$-\frac{1}{\rho_0}\operatorname{div}\left(\frac{1}{\theta}\boldsymbol{q}\right)+\frac{1}{\theta}r=\frac{1}{\theta}\dot{e}-\boldsymbol{\Delta}\cdot\dot{\boldsymbol{E}}-\frac{\kappa}{\rho_0\theta^2}\boldsymbol{g}\cdot\boldsymbol{g}\leqslant\frac{1}{\theta}\dot{e}-\boldsymbol{\Delta}\cdot\dot{\boldsymbol{E}}\,.$$

In any cyclic process starting from equilibrium, defined as in section 2.2, we have, if the times t_0 and t_1 bear the same meaning as they do in that section, that $e(t_0)=e(t_1)$ and hence that

$$\int_{t_0}^{t_1}\frac{1}{\theta(e(t))}\dot{e}(t)\,dt=0$$

and that

$$\int_{t_0}^{t_1}\left\{-\frac{1}{\rho_0}\operatorname{div}\left(\frac{1}{\theta}\boldsymbol{q}\right)+\frac{1}{\theta}r\right\}dt\leqslant-\int_{t_0}^{t_1}\boldsymbol{\Delta}(t,\boldsymbol{E}(\cdot))\cdot\dot{\boldsymbol{E}}(t)\,dt\,.$$

accordingly the thermodynamic inequality (2.2.3) holds provided that

$$\int_{t_0}^{t_1}\boldsymbol{\Delta}(t,\boldsymbol{E}(\cdot))\cdot\dot{\boldsymbol{E}}(t)\,dt\geqslant 0 \tag{4.2.16}$$

[1] We have replaced ρ by ρ_0.

around every closed path $\boldsymbol{E}(\cdot)$ in strain space that is a path for which $\boldsymbol{E}(t)$ is equal to some constant symmetric tensor for all times $t \leqslant t_0$ and is equal to the same tensor for all times $t \geqslant t_1$. Naturally the inequality (4.2.16) restricts the behaviour of the relaxation $\mathscr{G}(\cdot)$ considerably; we shall assume from now on that it does have this property. As we shall see the isotropic relaxation function (4.2.12) meets all our requirements if $\lambda(\cdot)$ and $\mu(\cdot)$ are given by finite sums

$$\lambda(t)=\sum_{n=1}^{N} \lambda_n \exp(-p_n t), \qquad \mu(t)=\sum_{n=1}^{N} \mu_n \exp(-p_n t) \tag{4.2.17}$$

of exponential functions with non-negative constants λ_n, μ_n and p_n.

We turn to verifying that the fading memory assumption holds. The Clausius integral taken between any two times τ_0 and τ_1 is

$$\begin{aligned}\int_{\tau_0}^{\tau_1} \frac{1}{\theta(t)} h(t)\,dt &= \int_{\tau_0}^{\tau_1} \left(\frac{1}{\theta(e(t))}\dot{e}(t) - \boldsymbol{\Delta}(t)\cdot\dot{\boldsymbol{E}}(t)\right) dt \\ &= \Theta(e(\tau_1)) - \Theta(e(\tau_0)) - \int_{\tau_0}^{\tau_1} \boldsymbol{\Delta}(t)\cdot\dot{\boldsymbol{E}}(t)\,dt \end{aligned} \tag{4.2.18}$$

where $\Theta(\cdot)$ has the derivative $(d/de)\Theta(e)=1/\theta(e)$. Suppose that $\boldsymbol{E}_1(\cdot)$ and $\boldsymbol{E}_2(\cdot)$ are any two paths in strain space with the final value of the first coinciding with the initial value of the second; that is

$$\boldsymbol{E}_1(+\infty)=\boldsymbol{E}_2(-\infty).$$

If we choose times u_1 and u_2 with $\boldsymbol{E}_1(t)=\boldsymbol{E}_1(+\infty)$ for every $t \geqslant u_1$ and $\boldsymbol{E}_2(t)=\boldsymbol{E}_2(-\infty)$ for every $t \leqslant u_2$ and form the composite paths $\boldsymbol{E}_\beta(\cdot)$ by setting $\boldsymbol{E}_\beta(t)=\boldsymbol{E}_1(t+\beta)$ for $t \leqslant u_2$ and $\boldsymbol{E}_\beta(t)=\boldsymbol{E}_2(t)$ for $t>u_2$ then, because of (4.2.18), all we have to verify to ensure that the material has fading memory in the sense of section 3.2 is that

$$\int_{-\infty}^{+\infty} \boldsymbol{\Delta}(t,\boldsymbol{E}_\beta(\cdot))\cdot\boldsymbol{E}_\beta(t)\,dt \to \int_{-\infty}^{+\infty} \boldsymbol{\Delta}(t,\boldsymbol{E}_1(\cdot))\cdot\dot{\boldsymbol{E}}_1(t)\,dt + \int_{-\infty}^{+\infty} \boldsymbol{\Delta}(t,\boldsymbol{E}_2(\cdot))\cdot\dot{\boldsymbol{E}}_2(t)\,dt \tag{4.2.19}$$

as $\beta \to +\infty$. Because the path $\boldsymbol{E}_\beta(\cdot)$ is constant on the interval $u_1-\beta \leqslant t \leqslant u_2$

$$\begin{aligned}\int_{-\infty}^{+\infty} \boldsymbol{\Delta}(t,\boldsymbol{E}_\beta(\cdot))\cdot\dot{\boldsymbol{E}}_\beta(t)\,dt &= \left(\int_{-\infty}^{u_1-\beta} + \int_{u_2}^{+\infty}\right) \boldsymbol{\Delta}(t,\boldsymbol{E}_\beta(\cdot))\cdot\dot{\boldsymbol{E}}_\beta(t)\,dt \\ &= \int_{-\infty}^{+\infty} \boldsymbol{\Delta}(t,\boldsymbol{E}(\cdot))\cdot\dot{\boldsymbol{E}}_1(t)\,dt \\ &\quad + \int_{u_2}^{+\infty} \boldsymbol{\Delta}(t,\boldsymbol{E}_\beta(\cdot))\cdot\dot{\boldsymbol{E}}_2(t)\,dt\,. \end{aligned} \tag{4.2.20}$$

For times $t \geqslant u_2$

$$
\begin{aligned}
\Delta(t, \boldsymbol{E}_\beta(\cdot)) &= \mathcal{G}(0)\,\boldsymbol{E}_\beta(t) + \int_{-\infty}^{t} \dot{\mathcal{G}}(t-s)\,\boldsymbol{E}_\beta(s)\,ds \\
&= \mathcal{G}(0)\,\boldsymbol{E}_2(t) + \left(\int_{-\infty}^{u_1-\beta} + \int_{u_1-\beta}^{t}\right)\dot{\mathcal{G}}(t-s)\,\boldsymbol{E}_\beta(s)\,ds \\
&= \mathcal{G}(0)\,\boldsymbol{E}_2(t) + \int_{u_1-\beta}^{t} \dot{\mathcal{G}}(t-s)\,\boldsymbol{E}_2(s)\,ds + \int_{-\infty}^{u_1-\beta} \dot{\mathcal{G}}(t-s)\,\boldsymbol{E}_1(s+\beta)\,ds \\
&= \Delta(t, \boldsymbol{E}_2(\cdot)) + \int_{t-u_1+\beta}^{+\infty} \dot{\mathcal{G}}(s)\{\boldsymbol{E}_1(t-s+\beta) - \boldsymbol{E}_2(t-s)\}\,ds \qquad (4.2.21)
\end{aligned}
$$

and so, bearing in mind that $\|\boldsymbol{E}_1(\cdot)\|$ and $\|\boldsymbol{E}_2(\cdot)\|$ are bounded and that $\dot{\boldsymbol{E}}_1(\cdot)$ and $\dot{\boldsymbol{E}}_2(\cdot)$ vanish outside an interval of finite length, it follows from equations (4.2.20) and (4.2.21) that if the relaxation function has an absolutely integrable derivative in the sense that

$$
\int_{0}^{+\infty} \|\dot{\mathcal{G}}(s)\|\,ds < +\infty \qquad (4.2.22)
$$

then

$$
\left|\int_{-\infty}^{+\infty} \Delta(t, \boldsymbol{E}_\beta(\cdot))\cdot\dot{\boldsymbol{E}}_\beta(t)\,dt - \int_{-\infty}^{+\infty} \Delta(t, \boldsymbol{E}_1(\cdot))\cdot\dot{\boldsymbol{E}}_1(t)\,dt - \int_{-\infty}^{+\infty} \Delta(t, \boldsymbol{E}_2)\cdot\dot{\boldsymbol{E}}_2(t)\,dt\right|
$$
$$
\leqslant C(\boldsymbol{E}_1(\cdot)), \boldsymbol{E}_2(\cdot)) \int_{u_2-u_1+\beta}^{+\infty} \|\dot{\mathcal{G}}(s)\|\,ds\,,
$$

where $C(\boldsymbol{E}_1(\cdot), \boldsymbol{E}_2(\cdot))$ is a constant depending only on the paths $\boldsymbol{E}_1(\cdot)$ and $\boldsymbol{E}_2(\cdot)$. On letting $\beta \to +\infty$ we deduce the required result (4.2.19) and so the material has fading memory provided the condition (4.2.22) holds.

Next we check that the material behaves suitably under retardation. Because of equation (4.2.18) all we need check in that if $\boldsymbol{E}(\cdot)$ is any path in strain space which is constant before t_0 and after t_1 and if $\boldsymbol{E}(\cdot, \alpha)$ is the retarded path

$$
\boldsymbol{E}(t, \alpha) = \boldsymbol{E}(t_0 + \alpha(t - t_0)), \qquad -\infty < t < +\infty, \qquad (4.2.23)
$$

then

$$
\int_{-\infty}^{+\infty} \Delta(t, \boldsymbol{E}(\cdot, \alpha))\cdot\dot{\boldsymbol{E}}(t, \alpha)\,dt \to \int_{-\infty}^{+\infty} \Delta^*(\boldsymbol{E}(t))\cdot\dot{\boldsymbol{E}}(t)\,dt \qquad (4.2.24)
$$

as $\alpha \to 0$, where $\Delta^*(\cdot)$ is defined in equation (4.2.15). Now

$$
\int_{-\infty}^{+\infty} \Delta(t, \boldsymbol{E}(\cdot, \alpha))\cdot\dot{\boldsymbol{E}}(t, \alpha)\,dt = \int_{t_0}^{+\infty}\left\{\mathcal{G}(0)\,\boldsymbol{E}(t, \alpha) + \int_{0}^{+\infty} \dot{\mathcal{G}}(s)\,\boldsymbol{E}(t-s, \alpha)\,ds\right\}\cdot\dot{\boldsymbol{E}}(t, \alpha)\,dt
$$

and on making the change of variable $u=t_0+\alpha(t-t_0)$ and using the definition (4.2.23) we find that

$$\begin{aligned}&\int_{-\infty}^{+\infty} \Delta(t,\boldsymbol{E}(\cdot,\alpha))\cdot\boldsymbol{E}(t,\alpha)\,dt\\ &\quad=\int_{t_0}^{t_1}\left\{\mathscr{G}(0)\boldsymbol{E}(u)+\int_0^{+\infty}\dot{\mathscr{G}}(s)\boldsymbol{E}(u-\alpha s)\,ds\right\}\cdot\dot{\boldsymbol{E}}(u)\,du\\ &\quad=\int_{t_0}^{t_1}\mathscr{G}(+\infty)\boldsymbol{E}(u)\cdot\dot{\boldsymbol{E}}(u)\,du+\int_{t_0}^{t_1}\int_0^{+\infty}\dot{\mathscr{G}}(s)\{\boldsymbol{E}(u-\alpha s)-\boldsymbol{E}(u)\}\,ds\,du.\end{aligned}$$

The fact that the norm $|\dot{\boldsymbol{E}}(\cdot)|$ of the strain rate is a bounded function implies, together with the mean-value theorem, the estimate

$$|\boldsymbol{E}(u-\alpha s)-\boldsymbol{E}(u)|\leqslant c(\boldsymbol{E}(\cdot))\alpha s,$$

in which $c(\boldsymbol{E}(\cdot))$ is a constant depending only on the path $\boldsymbol{E}(\cdot)$, and consequently

$$\begin{aligned}&\int_{-\infty}^{+\infty}\Delta(t,\boldsymbol{E}(\cdot,\alpha))\cdot\dot{\boldsymbol{E}}(t,\alpha)\,dt-\int_{-\infty}^{+\infty}\Delta^*(\boldsymbol{E}(t))\cdot\dot{\boldsymbol{E}}(t)\,dt\\ &\qquad\leqslant\alpha(t_1-t_0)c(\boldsymbol{E}(\cdot))\int_0^{+\infty}s\|\dot{\mathscr{G}}(s)\|\,ds\end{aligned}\tag{4.2.25}$$

provided that the relaxation function satisfies

$$\int_0^{+\infty}s\|\dot{\mathscr{G}}(s)\|\,ds<+\infty.\tag{4.2.26}$$

On letting $\alpha\to 0$ in (4.2.25) we deduce the relation (4.2.24) and hence the material is well-behaved under retardation.

In summary, the material defined by the constitutive relations (4.2.4), (4.2.5) and (4.2.6) meets all our requirements provided that the inequality (4.2.16) holds for every closed path in strain space and provided that the functions $\dot{\mathscr{G}}(s)$ and $s\dot{\mathscr{G}}(s)$ are absolutely integrable. This means that the thermostatic relation (3.3.10) must hold; for this material it states that

$$\left(\frac{\partial\eta^*}{\partial\boldsymbol{E}},\frac{\partial\eta^*}{\partial e}\right)=\left(-\Delta^*,\frac{1}{\theta}\right)\tag{4.2.27}$$

and so the *equilibrium entropy* is, to within an arbitrary additive constant,

$$\eta^*(\boldsymbol{E},e)=-\tfrac{1}{2}\boldsymbol{E}\cdot\mathscr{G}(+\infty)\boldsymbol{E}+\Theta(e).\tag{4.2.28}$$

To compute the entropy we must first find the recoverable entropy. The Clausius integral (4.2.18) reduces if $e(\tau_0)=e(\tau_1)$ to

$$\int_{\tau_0}^{\tau_1}\frac{h(t)}{\theta(t)}\,dt=-\int_{\tau_0}^{\tau_1}\Delta(t,\boldsymbol{E}(\cdot))\cdot\dot{\boldsymbol{E}}(t)\,dt$$

which means that the entropy $\overset{\tau}{\underset{-\infty}{H}}(\boldsymbol{E}(\cdot), e(\cdot))$ recoverable from a path $\boldsymbol{E}(\cdot)$ at time τ is the least upper bound of the integrals

$$-\int_{\tau}^{+\infty} \Delta(t, \boldsymbol{A}(\cdot)) \cdot \dot{\boldsymbol{A}}(t)\, dt \tag{4.2.29}$$

as the path $\boldsymbol{A}(\cdot)$ varies over all closed connections of $\boldsymbol{E}(\cdot)$ at τ, that is $\boldsymbol{A}(t) \equiv \boldsymbol{E}(t)$ for all $t \leqslant \tau$ and the final value of $\boldsymbol{A}(t)$ is $\boldsymbol{A}(+\infty) = \boldsymbol{A}(\tau) = \boldsymbol{E}(\tau)$. We shall determine the value of this least upper bound and incidentally our investigation will tell us not just its value but it will also produce closed connections $\boldsymbol{A}(\cdot)$ for which the integral (4.2.29) approaches the recoverable entropy as closely as we wish.

The evaluation of the recoverable entropy and the construction of closed connections maximising the integrals (4.2.29) is intimately connected with a concept of reversibility. Suppose that $\boldsymbol{E}_{\text{rev}}(\cdot)$ is a function whose values are symmetric strain tensors with $\boldsymbol{E}_{\text{rev}}(t) \equiv \boldsymbol{E}(t)$ for every $t \leqslant \tau$. We allow $\boldsymbol{E}_{\text{rev}}(\cdot)$ to have a jump discontinuity at τ, that is $\lim_{t \to \tau+} \boldsymbol{E}_{\text{rev}}(t) \neq \boldsymbol{E}(\tau)$, but it must be continuous everywhere else. We can regard $\boldsymbol{E}_{\text{rev}}(t)$ as the strain at time t in a process in the material. Let $e_{\text{rev}}(t)$ be the internal energy at time t in this process, then the *time reversal* of the process about the time τ is that process in which the strain and the internal energy run through their values in the reverse order, that is to say the strain is $\overline{\boldsymbol{E}}_{\text{rev}}(t) = \boldsymbol{E}_{\text{rev}}(2\tau - t)$ and the internal energy is $\overline{e}_{\text{rev}}(t) = e_{\text{rev}}(2\tau - t)$, for every t in $-\infty < t < +\infty$. Because of equation (4.2.5) the temperature in the reversed process is $\theta(\overline{e}_{\text{rev}}(t)) = \theta(e_{\text{rev}}(2\tau - t))$. In general, because of the memory of the material, the stress does not satisfy

$$\boldsymbol{T}(t, \boldsymbol{E}_{\text{rev}}(\cdot), e_{\text{rev}}(\cdot)) = \boldsymbol{T}(2\tau - t, \overline{\boldsymbol{E}}_{\text{rev}}(\cdot), \overline{e}_{\text{rev}}(\cdot)), \qquad t > \tau, \tag{4.2.30}$$

but if it does do so for every $t > \tau$ we shall say that $\boldsymbol{E}_{\text{rev}}(\cdot)$ is a *reversible extension* of the path $\boldsymbol{E}(\cdot)$ at τ. The necessary and sufficient condition for $\boldsymbol{E}_{\text{rev}}(\cdot)$ to be a reversible extension of $\boldsymbol{E}(\cdot)$ at τ is that it shall satisfy the integral equation[1]

$$\int_{\tau}^{+\infty} \left(\frac{d}{ds} \mathscr{G}(|t-s|) \right) \boldsymbol{E}_{\text{rev}}(s)\, ds = \int_{-\infty}^{\tau} \dot{\mathscr{G}}(t-s) \boldsymbol{E}(s)\, ds \tag{4.2.31}$$

for every $t > \tau$. This is an equation of the Wiener-Hopf kind.

For the validity of the arguments which follow we shall need to assume that the path $\boldsymbol{E}(\cdot)$ and the relaxation function $\mathscr{G}(\cdot)$ are such that

---

[1] An integral equation equivalent to (4.2.31) was first derived, using a variational argument, by Breuer and Onat [2] in their study of recoverable work. The argument which is used here is that of Day [34].

equation (4.2.31) has a solution $\boldsymbol{E}_{\mathrm{rev}}(\cdot)$ which is continuous on $\tau<t<+\infty$. and such that the limits $\boldsymbol{E}_{\mathrm{rev}}(\tau)=\lim_{t\to\tau+}\boldsymbol{E}_{\mathrm{rev}}(t)$ and $\boldsymbol{E}_{\mathrm{rev}}(+\infty)=\lim_{t\to+\infty}\boldsymbol{E}_{\mathrm{rev}}(t)$ exist. Furthermore we shall require that $\boldsymbol{E}_{\mathrm{rev}}(t)$ approaches its limiting value $\boldsymbol{E}_{\mathrm{rev}}(+\infty)$ for large t sufficiently rapidly for the integrals

$$\int_{\tau}^{+\infty} |\boldsymbol{E}_{\mathrm{rev}}(t)-\boldsymbol{E}_{\mathrm{rev}}(+\infty)|\,dt$$

and

$$\int_{\tau}^{+\infty} |\boldsymbol{E}_{\mathrm{rev}}(t)-\boldsymbol{E}_{\mathrm{rev}}(+\infty)|^2\,dt$$

to be finite.

Let the path $\boldsymbol{A}(\cdot)$ be any closed connection of the path $\boldsymbol{E}(\cdot)$ at τ so that $\boldsymbol{A}(+\infty)=\boldsymbol{A}(\tau)=\boldsymbol{E}(\tau)$. An integration by parts shows that the integral (4.2.29) can be written as

$$\int_{\tau}^{+\infty} \dot{\boldsymbol{\Delta}}(t,\boldsymbol{A}(\cdot))\cdot(\boldsymbol{A}(t)-\boldsymbol{A}(\tau))\,dt$$

and since, for $t>\tau$,

$$\begin{aligned}\boldsymbol{\Delta}(t,\boldsymbol{A}(\cdot))&=\mathscr{G}(0)\boldsymbol{A}(t)+\int_{-\infty}^{t}\dot{\mathscr{G}}(t-s)\boldsymbol{A}(s)\,ds\\&=\mathscr{G}(0)\boldsymbol{A}(t)+\int_{\tau}^{t}\dot{\mathscr{G}}(t-s)(\boldsymbol{A}(s)-\boldsymbol{A}(\tau))\,ds\\&\quad+\int_{-\infty}^{\tau}\dot{\mathscr{G}}(t-s)(\boldsymbol{E}(s)-\boldsymbol{E}(\tau))\,ds-(\mathscr{G}(0)-\mathscr{G}(+\infty))\boldsymbol{E}(\tau)\end{aligned}$$

we have

$$-\int_{\tau}^{+\infty}\boldsymbol{\Delta}(t,\boldsymbol{A}(\cdot))\cdot\dot{\boldsymbol{A}}(t)\,dt=\Phi_{\tau}(\boldsymbol{E}(\cdot),\boldsymbol{A}(\cdot)-\boldsymbol{A}(\tau))\tag{4.2.32}$$

where the scalar functional $\Phi_{\tau}(\cdot)$ is defined by

$$\begin{aligned}\Phi_{\tau}(\boldsymbol{B}_1(\cdot),\boldsymbol{B}_2(\cdot))&=\int_{\tau}^{+\infty}\int_{-\infty}^{\tau}\boldsymbol{B}_2(t)\cdot\ddot{\mathscr{G}}(t-s)(\boldsymbol{B}_1(s)-\boldsymbol{B}_1(\tau))\,ds\,dt\\&\quad+\int_{\tau}^{+\infty}\boldsymbol{B}_2(t)\cdot\dot{\mathscr{G}}(0)\boldsymbol{B}_2(t)\,dt\\&\quad+\int_{\tau}^{+\infty}\int_{\tau}^{t}\boldsymbol{B}_2(t)\cdot\ddot{\mathscr{G}}(t-s)\boldsymbol{B}_2(s)\,ds\,dt\end{aligned}\tag{4.2.33}$$

whenever the integrals exist.

If $\boldsymbol{E}_{\mathrm{rev}}(\cdot)$ is a reversible extension of $\boldsymbol{E}(\cdot)$ at τ and if $\boldsymbol{C}(\cdot)$ is defined for $t>\tau$ by

$$\boldsymbol{A}(t)-\boldsymbol{A}(\tau)=\boldsymbol{E}_{\mathrm{rev}}(t)-\boldsymbol{E}_{\mathrm{rev}}(+\infty)+\boldsymbol{C}(t)\tag{4.2.34}$$

then $C(\tau)=E_{rev}(+\infty)-E_{rev}(\tau)$ and $C(t)\to O$ as $t\to+\infty$ and the integrals $\int_\tau^{+\infty}|C(t)|\,dt$ and $\int_\tau^{+\infty}|C(t)|^2\,dt$ are finite. On substituting from (4.2.34) into (4.2.32) we find that

$$\begin{aligned}
-\int_\tau^{+\infty}\Delta(t,A(\cdot))\cdot\dot{A}(t)\,dt &= \Phi_\tau(E(\cdot),E_{rev}(\cdot)-E_{rev}(+\infty))+\Phi_\tau(E(\cdot),C(\cdot))\\
&\quad+2\int_\tau^{+\infty}C(t)\cdot\dot{\mathscr{G}}(0)(E_{rev}(t)-E_{rev}(+\infty))\,dt\\
&\quad+\int_\tau^{+\infty}\int_\tau^{t}C(t)\cdot\ddot{\mathscr{G}}(t-s)(E_{rev}(s)-E_{rev}(+\infty))\,ds\,dt\\
&\quad+\int_\tau^{+\infty}\int_\tau^{t}C(s)\cdot\ddot{\mathscr{G}}(t-s)(E_{rev}(t)-E_{rev}(+\infty))\,ds\,dt.
\end{aligned}$$

However,

$$\begin{aligned}
&\int_\tau^{+\infty}\int_\tau^{t}C(s)\cdot\ddot{\mathscr{G}}(t-s)(E_{rev}(t)-E_{rev}(+\infty))\,ds\,dt\\
&\quad=\int_\tau^{+\infty}\int_t^{+\infty}C(t)\cdot\ddot{\mathscr{G}}(s-t)(E_{rev}(s)-E_{rev}(+\infty))\,ds\,dt
\end{aligned}$$

and if we use the definition (4.2.33) to evaluate the functional $\Phi_\tau(E(\cdot),C(\cdot))$ we obtain the expression

$$\begin{aligned}
-\int_\tau^{+\infty}\Delta(t,A)\cdot\dot{A}(t)\,dt &= \Phi_\tau(E(\cdot),E_{rev}(\cdot)-E_{rev}(+\infty))\\
&\quad+\int_\tau^{+\infty}C(t)\cdot\dot{\mathscr{G}}(0)C(t)\,dt\\
&\quad+\int_\tau^{+\infty}\int_\tau^{t}C(t)\cdot\ddot{\mathscr{G}}(t-s)C(s)\,ds\,dt\\
&\quad+\int_\tau^{+\infty}C(t)\cdot\Big\{2\dot{\mathscr{G}}(0)(E_{rev}(t)-E_{rev}(+\infty))\\
&\quad+\int_\tau^{+\infty}\ddot{\mathscr{G}}(|t-s|)(E_{rev}(s)-E_{rev}(+\infty))\,ds\\
&\quad+\int_{-\infty}^{\tau}\ddot{\mathscr{G}}(t-s)(E(s)-E(\tau))\,ds\Big\}\,dt. \qquad (4.2.35)
\end{aligned}$$

The reversible extension $E_{rev}(\cdot)$ satisfies equation (4.2.31). Differentiating this equation throughout with respect to t yields the equation

$$2\dot{\mathscr{G}}(0)E_{rev}(t)+\int_\tau^{+\infty}\ddot{\mathscr{G}}(|t-s|)E_{rev}(s)\,ds+\int_{-\infty}^{\tau}\ddot{\mathscr{G}}(t-s)E(s)\,ds=0$$

which means that (4.2.35) simplifies to

$$-\int_{\tau}^{+\infty} \Delta(t,\boldsymbol{A})\cdot \boldsymbol{A}(t)\,dt = \Phi_\tau(\boldsymbol{E}(\cdot),\boldsymbol{E}_{\text{rev}}(\cdot)-\boldsymbol{E}_{\text{rev}}(+\infty)) + \int_{\tau}^{+\infty} \boldsymbol{C}(t)\cdot\dot{\mathscr{G}}(0)\boldsymbol{C}(t)\,dt$$
$$+\int_{\tau}^{+\infty}\int_{\tau}^{t} \boldsymbol{C}(t)\cdot\ddot{\mathscr{G}}(t-s)\boldsymbol{C}(s)\,ds\,dt$$
$$+(\boldsymbol{E}(\tau)-\boldsymbol{E}_{\text{rev}}(+\infty))\cdot\int_{\tau}^{+\infty}\dot{\mathscr{G}}(t-\tau)\boldsymbol{C}(t)\,dt \qquad (4.2.36)$$

and it is this expression we shall use to find the recoverable entropy. In fact we shall show that the recoverable entropy is

$$\begin{aligned}&\Phi_\tau(\boldsymbol{E}(\cdot),\boldsymbol{E}_{\text{rev}}(\cdot)-\boldsymbol{E}_{\text{rev}}(+\infty))\\ &\quad+\tfrac{1}{2}(\boldsymbol{E}(\tau)-\boldsymbol{E}_{\text{rev}}(+\infty))\cdot(\mathscr{G}(0)-\mathscr{G}(+\infty))(\boldsymbol{E}(\tau)-\boldsymbol{E}_{\text{rev}}(+\infty)).\end{aligned} \qquad (4.2.37)$$

To do this we begin by showing that the expression (4.2.37) is indeed an upper bound, that is to say

$$\int_{\tau}^{+\infty} \boldsymbol{C}(t)\cdot\dot{\mathscr{G}}(0)\boldsymbol{C}(t)\,dt + \int_{\tau}^{+\infty}\int_{\tau}^{t} \boldsymbol{C}(t)\cdot\ddot{\mathscr{G}}(t-s)\boldsymbol{C}(s)\,ds\,dt$$
$$+(\boldsymbol{E}(\tau)-\boldsymbol{E}_{\text{rev}}(+\infty))\cdot\int_{\tau}^{+\infty}\dot{\mathscr{G}}(t-\tau)\boldsymbol{C}(t)\,dt$$
$$\leqslant \tfrac{1}{2}(\boldsymbol{E}(\tau)-\boldsymbol{E}_{\text{rev}}(+\infty))\cdot(\mathscr{G}(0)-\mathscr{G}(+\infty))(\boldsymbol{E}(\tau)-\boldsymbol{E}_{\text{rev}}(+\infty)). \qquad (4.2.38)$$

Let n_1 and n_2 be any large positive integers and let $\boldsymbol{C}_{n_1 n_2}(\cdot)$ be the double sequence of strain paths defined by setting

$$\boldsymbol{C}_{n_1 n_2}(t) = \begin{cases} \boldsymbol{E}_{\text{rev}}(+\infty)-\boldsymbol{E}(\tau), & t\leqslant \tau-n_1^{-1};\\ \boldsymbol{E}_{\text{rev}}(+\infty)-\boldsymbol{E}(\tau)+n_1(t-\tau+n_1^{-1})(\boldsymbol{C}(\tau) & -\boldsymbol{E}_{\text{rev}}(+\infty)+\boldsymbol{E}(\tau)),\\ & \tau-n_1^{-1}\leqslant t\leqslant\tau;\\ \boldsymbol{C}(t), & \tau\leqslant t\leqslant \tau+n_2;\\ \boldsymbol{C}(\tau+n_2), & \tau+n_2\leqslant t\,. \end{cases} \qquad (4.2.39)$$

The behaviour of this double sequence is indicated in Fig. 5. Since $\boldsymbol{C}_{n_1 n_2}(-\infty)=\boldsymbol{E}_{\text{rev}}(+\infty)-\boldsymbol{E}(\tau)$ and $\boldsymbol{C}_{n_1 n_2}(+\infty)=\boldsymbol{C}(\tau+n_2)$ the inequality (3.3.14), proved in section 3.3, and the expression (4.2.28) for the equilibrium entropy together tell us that

$$\tfrac{1}{2}(\boldsymbol{E}_{\text{rev}}(+\infty)-\boldsymbol{E}(\tau))\cdot\mathscr{G}(+\infty)(\boldsymbol{E}_{\text{rev}}(+\infty)-\boldsymbol{E}(\tau)$$
$$-\tfrac{1}{2}\boldsymbol{C}(\tau+n_2)\cdot\mathscr{G}(+\infty)\boldsymbol{C}(\tau+n_2) \geqslant -\int_{-\infty}^{+\infty}\Delta(t,\boldsymbol{C}_{n_1 n_2})\cdot\dot{\boldsymbol{C}}_{n_1 n_2}(t)\,dt$$
$$= -\left(\int_{\tau-n_1^{-1}}^{\tau} + \int_{\tau}^{\tau+n_2}\right)\Delta(t,\boldsymbol{C}_{n_1 n_2})\cdot\dot{\boldsymbol{C}}_{n_1 n_2}(t)\,dt\,. \qquad (4.2.40)$$

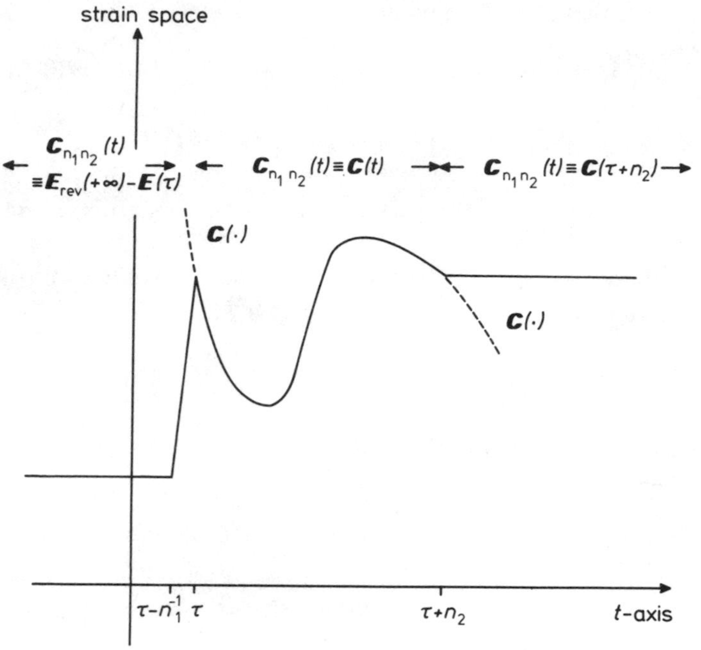

Fig. 5: The double sequence of paths $C_{n_1n_2}(\cdot)$

But

$$\int_{\tau-n_1^{-1}}^{\tau} \Delta(t, C_{n_1n_2}(\cdot))\cdot \dot{C}_{n_1n_2}(t)\,dt$$

$$= \int_{\tau-n_1^{-1}}^{\tau} \Big\{\mathscr{G}(0)\,C_{n_1n_2}(t)+(\mathscr{G}(+\infty)-\mathscr{G}(0))(E_{\text{rev}}(+\infty)-E(\tau))$$

$$+\int_{\tau-n_1^{-1}}^{t} \dot{\mathscr{G}}(t-s)\,C_{n_1n_2}(s)\,ds\Big\}\cdot \dot{C}_{n_1n_2}(t)\,dt$$

from which we can deduce that

$$\lim_{n_2\to+\infty}\ \lim_{n_1\to+\infty}\ \int_{\tau-n_1^{-1}}^{\tau} \Delta(t, C_{n_1n_2}(\cdot))\cdot \dot{C}_{n_1n_2}(t)\,dt$$

$$=\tfrac{1}{2}C(\tau)\cdot\mathscr{G}(0)\,C(\tau)-\tfrac{1}{2}(E_{\text{rev}}(+\infty)-E(\tau))\cdot\mathscr{G}(0)\,(E_{\text{rev}}(+\infty)-E(\tau))$$

$$+(C(\tau)-E_{\text{rev}}(+\infty)+E(\tau))\cdot(\mathscr{G}(+\infty)-\mathscr{G}(0))(E_{\text{rev}}(+\infty)-E(\tau)). \tag{4.2.41}$$

Also, an integration by parts shows that

$$\begin{aligned}\int_{\tau}^{\tau+n_2} & \boldsymbol{\Delta}(t,\boldsymbol{C}_{n_1n_2}(\cdot))\cdot\dot{\boldsymbol{C}}_{n_1n_2}(t)\,dt\\ &= \boldsymbol{C}(\tau+n_2)\cdot\boldsymbol{\Delta}(\tau+n_2,\boldsymbol{C}_{n_1n_2}(\cdot)) - \boldsymbol{C}(\tau)\cdot\boldsymbol{\Delta}(\tau,\boldsymbol{C}_{n_1n_2}(\cdot))\\ &\quad - \int_{\tau}^{\tau+n_2} \dot{\boldsymbol{\Delta}}(t,\boldsymbol{C}_{n_1n_2})\cdot\boldsymbol{C}(t)\,dt\end{aligned}$$

and thus, because of the fact that $\boldsymbol{C}(\tau+n_2)\to\boldsymbol{O}$ as $n_2\to+\infty$, we have

$$\begin{aligned}\lim_{n_2\to+\infty}\lim_{n_1\to+\infty} & \int_{\tau}^{\tau+n_2} \boldsymbol{\Delta}(t,\boldsymbol{C}_{n_1n_2}(\cdot))\cdot\dot{\boldsymbol{C}}_{n_1n_2}(t)\,dt\\ &= -\boldsymbol{C}(\tau)\cdot\{\mathcal{G}(0)\boldsymbol{C}(\tau)+(\mathcal{G}(+\infty)-\mathcal{G}(0))(\boldsymbol{E}_{\mathrm{rev}}(+\infty)-\boldsymbol{E}(\tau))\}\\ &\quad +\tfrac{1}{2}\boldsymbol{C}(\tau)\cdot\mathcal{G}(0)\boldsymbol{C}(\tau) - \int_{\tau}^{+\infty}\boldsymbol{C}(t)\cdot\dot{\mathcal{G}}(0)\boldsymbol{C}(t)\,dt\\ &\quad -\int_{\tau}^{+\infty}\int_{\tau}^{t}\boldsymbol{C}(t)\cdot\ddot{\mathcal{G}}(t-s)\boldsymbol{C}(s)\,ds\,dt\\ &\quad -(\boldsymbol{E}(\tau)-\boldsymbol{E}_{\mathrm{rev}}(+\infty))\cdot\int_{\tau}^{+\infty}\dot{\mathcal{G}}(t-\tau)\boldsymbol{C}(t)\,dt\,. \end{aligned}\tag{4.2.42}$$

If we now take the double limit $\lim_{n_2\to+\infty}\lim_{n_1\to+\infty}$ on both sides of (4.2.40) and use (4.2.42) we find, after some rearrangement of the terms, precisely the required inequality (4.2.38) and so (4.2.37) is indeed an upper bound for the recoverable entropy.

To show that the expression (4.2.37) is the recoverable entropy we must produce closed connections $\boldsymbol{A}(\cdot)$ of the path $\boldsymbol{E}(\cdot)$ at τ for which the integral

$$-\int_{\tau}^{+\infty}\boldsymbol{\Delta}(t,\boldsymbol{A}(\cdot))\cdot\dot{\boldsymbol{A}}(t)\,dt$$

approaches the expression as closely as we please. Let n_1, n_2, n_3 and n_4 be any large integers and define a multiple sequence $\boldsymbol{A}_{n_1n_2n_3n_4}(\cdot)$ of closed connections of $\boldsymbol{E}(\cdot)$ at τ by setting

$$\boldsymbol{A}_{n_1n_2n_3n_4}(t)=\begin{cases}\boldsymbol{E}(t), & t\leqslant\tau;\\ \boldsymbol{E}(\tau)+n_1(t-\tau)(\boldsymbol{E}_{\mathrm{rev}}(\tau)-\boldsymbol{E}(\tau)), & \tau\leqslant t\leqslant\tau+n_1^{-1};\\ \boldsymbol{E}_{\mathrm{rev}}(t), & \tau+n_1^{-1}\leqslant t\leqslant\tau+n_2;\\ \boldsymbol{E}_{\mathrm{rev}}(\tau+n_2), & \tau+n_2\leqslant t\leqslant\tau+n_3;\\ \boldsymbol{E}_{\mathrm{rev}}(\tau+n_2)+(n_4-n_3)^{-1}(t-\tau-n_3)(\boldsymbol{E}(\tau)-\boldsymbol{E}_{\mathrm{rev}}(\tau+n_2)), & \\ & \tau+n_3\leqslant t\leqslant\tau+n_4;\\ \boldsymbol{E}(\tau) & \tau+n_4\leqslant t\,.\end{cases}\tag{4.2.43}$$

The behaviour of this sequence is indicated in Fig. 6. It is not difficult to verify, by using arguments almost identical to those given already, that

$$\begin{aligned}
&\lim_{n_1\to+\infty}\ \lim_{n_4\to+\infty}\ \lim_{n_3\to+\infty}\ \int_{\tau}^{+\infty} \Delta(t, \boldsymbol{A}_{n_1n_2n_3n_4}(\cdot))\cdot \dot{\boldsymbol{A}}_{n_1n_2n_3n_4}(t)\,dt\\
&\quad=\tfrac{1}{2}\boldsymbol{E}_{\text{rev}}(\tau)\cdot\mathscr{G}(0)\,\boldsymbol{E}_{\text{rev}}(\tau)-\tfrac{1}{2}\boldsymbol{E}(\tau)\cdot\mathscr{G}(0)\,\boldsymbol{E}(\tau)\\
&\quad\quad+(\boldsymbol{E}_{\text{rev}}(\tau)-\boldsymbol{E}(\tau))\cdot\int_{-\infty}^{\tau}\dot{\mathscr{G}}(\tau-s)\,\boldsymbol{E}(s)\,ds\\
&\quad\quad+\tfrac{1}{2}\boldsymbol{E}(\tau)\cdot\mathscr{G}(+\infty)\,\boldsymbol{E}(\tau)-\tfrac{1}{2}\boldsymbol{E}_{\text{rev}}(\tau+n_2)\cdot\mathscr{G}(+\infty)\,\boldsymbol{E}_{\text{rev}}(\tau+n_2)\\
&\quad\quad+\int_{\tau}^{\tau+n_2}\boldsymbol{\Gamma}(t)\cdot\dot{\boldsymbol{E}}_{\text{rev}}(t)\,dt
\end{aligned}\tag{4.2.44}$$

where

$$\begin{aligned}
\boldsymbol{\Gamma}(t)=&\,\mathscr{G}(0)\,(\boldsymbol{E}_{\text{rev}}(t)-\boldsymbol{E}_{\text{rev}}(+\infty))\\
&+\int_{\tau}^{t}\dot{\mathscr{G}}(t-s)\,(\boldsymbol{E}_{\text{rev}}(s)-\boldsymbol{E}_{\text{rev}}(+\infty))\,ds\\
&+\int_{-\infty}^{\tau}\dot{\mathscr{G}}(t-s)\,(\boldsymbol{E}(s)-\boldsymbol{E}(\tau))\,ds\\
&+\dot{\mathscr{G}}(t-\tau)\,\{\boldsymbol{E}_{\text{rev}}(+\infty)-\boldsymbol{E}(\tau)\}+\mathscr{G}(+\infty)\,\boldsymbol{E}(\tau).
\end{aligned}\tag{4.2.45}$$

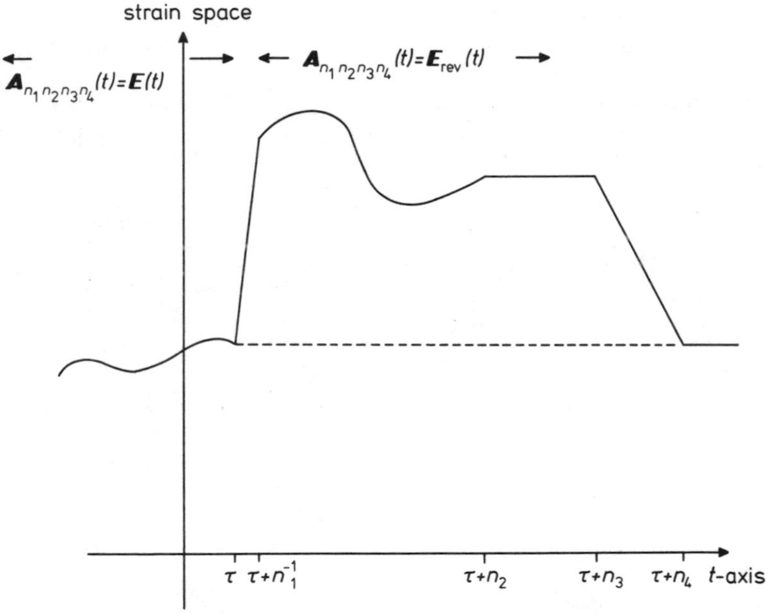

Fig. 6: The multiple sequence of closed connections $\boldsymbol{A}_{n_1n_2n_3n_4}(\cdot)$

If we integrate the second of the two integrals appearing on the right-hand side of (4.2.44) by parts and use (4.2.45) and afterwards take the limit as $n_2 \to \infty$ we find, after some rearrangement, that

$$\begin{aligned}
&\lim_{n_2\to+\infty}\lim_{n_1\to+\infty}\lim_{n_4\to+\infty}\lim_{n_3\to+\infty} - \int_\tau^{+\infty} \Delta(t, \boldsymbol{A}_{n_1n_2n_3n_4}(\cdot))\cdot \dot{\boldsymbol{A}}_{n_1n_2n_3n_4}(t)\,dt \\
&= \Phi_\tau(\boldsymbol{E}(\cdot), \boldsymbol{E}_{\text{rev}}(\cdot) - \boldsymbol{E}_{\text{rev}}(+\infty)) \\
&\quad - \tfrac{1}{2}(\boldsymbol{E}_{\text{rev}}(+\infty) - \boldsymbol{E}(\tau))\cdot(\mathscr{G}(0) - \mathscr{G}(+\infty))(\boldsymbol{E}_{\text{rev}}(+\infty) - \boldsymbol{E}(\tau)) \\
&\quad + (\boldsymbol{E}_{\text{rev}}(+\infty) - \boldsymbol{E}(\tau))\cdot\left(\int_\tau^{+\infty} \dot{\mathscr{G}}(s-\tau)(\boldsymbol{E}_{\text{rev}}(s) - \boldsymbol{E}_{\text{rev}}(+\infty))\,ds\right. \\
&\qquad\qquad \left. - \int_{-\infty}^{\tau} \dot{\mathscr{G}}(\tau - s)(\boldsymbol{E}(s) - \boldsymbol{E}(\tau))\,ds\right). \qquad (4.2.46)
\end{aligned}$$

However, if we let $t \to \tau$ in the integral equation (4.2.31) which determines the reversible extension $\boldsymbol{E}_{\text{rev}}(\cdot)$ we find that

$$\int_\tau^{+\infty} \dot{\mathscr{G}}(s-\tau)\,\boldsymbol{E}_{\text{rev}}(s)\,ds = \int_{-\infty}^{\tau} \dot{\mathscr{G}}(\tau - s)\,\boldsymbol{E}(s)\,ds$$

which implies that the multiple limit (4.2.46) does reduce to the expression (4.2.37) which must accordingly be the recoverable entropy. In other words we have shown that the recoverable entropy is

$$\begin{aligned}
\overset{\tau}{\underset{-\infty}{H}}(\boldsymbol{E}(\cdot), e(\cdot)) &= \Phi_\tau(\boldsymbol{E}(\cdot), \boldsymbol{E}_{\text{rev}}(\cdot) - \boldsymbol{E}_{\text{rev}}(+\infty)) \\
&\quad + \tfrac{1}{2}(\boldsymbol{E}(\tau) - \boldsymbol{E}_{\text{rev}}(+\infty))\cdot(\mathscr{G}(0) - \mathscr{G}(+\infty)(\boldsymbol{E}(\tau) - \boldsymbol{E}_{\text{rev}}(+\infty))
\end{aligned} \qquad (4.2.47)$$

and that the entropy is

$$\begin{aligned}
\overset{\tau}{\underset{-\infty}{\eta}}(\boldsymbol{E}(\cdot), e(\cdot)) &= \Theta(e(\tau)) - \tfrac{1}{2}\boldsymbol{E}(\tau)\cdot\mathscr{G}(+\infty)\boldsymbol{E}(\tau) \\
&\quad - \tfrac{1}{2}(\boldsymbol{E}(\tau) - \boldsymbol{E}_{\text{rev}}(+\infty))\cdot(\mathscr{G}(0) - \mathscr{G}(+\infty))(\boldsymbol{E}(\tau) - \boldsymbol{E}_{\text{rev}}(+\infty)) \\
&\quad - \Phi_\tau(\boldsymbol{E}(\cdot), \boldsymbol{E}_{\text{rev}}(\cdot) - \boldsymbol{E}_{\text{rev}}(+\infty))
\end{aligned} \qquad (4.2.48)$$

Moreover we have produced a multiple sequence of closed connections of $\boldsymbol{E}(\cdot)$ at τ, namely the sequence $\boldsymbol{A}_{n_1n_2n_3n_4}(\cdot)$ defined by (4.2.43), which ultimately extracts all the recoverable entropy. The pointwise limit of this sequence is

$$\lim_{n_2\to+\infty}\lim_{n_1\to+\infty}\lim_{n_4\to+\infty}\lim_{n_3\to+\infty} \boldsymbol{A}_{n_1n_2n_3n_4}(t) = \begin{cases} \boldsymbol{E}(t), & t \leqslant \tau, \\ \boldsymbol{E}_{\text{rev}}(t), & t > \tau, \end{cases} \qquad (4.2.49)$$

which, in general, has a jump discontinuity at $t = \tau$ and thereafter coincides with the reversible extension $\boldsymbol{E}_{\text{rev}}(\cdot)$ of $\boldsymbol{E}(\cdot)$.

As a concrete example let us consider the isotropic relaxation function (4.2.12) with the functions $\lambda(\cdot)$ and $\mu(\cdot)$ having the very special forms

$$\left.\begin{aligned}\lambda(s) &= \lambda(+\infty)+(\lambda(0)-\lambda(+\infty))\exp(-ks)\\ \mu(s) &= \mu(+\infty)+(\mu(0)-\mu(+\infty))\exp(-ks)\end{aligned}\right\} \tag{4.2.50}$$

where k is a positive constant. Clearly the integrals

$$\int_0^{+\infty} |\dot{\lambda}(s)|\,ds, \quad \int_0^{+\infty} s|\dot{\lambda}(s)|\,ds, \quad \int_0^{+\infty} |\dot{\mu}(s)|\,ds, \quad \int_0^{+\infty} s|\dot{\mu}(s)|\,ds$$

are all finite and the conditions (4.2.22) and (4.2.26) are met. We shall show too that provided

$$\lambda(0)>\lambda(+\infty), \qquad \mu(0)>\mu(+\infty) \tag{4.2.51}$$

the inequality (4.2.16) holds and that all our assumptions are satisfied.

It is convenient to introduce the functional $\tilde{\boldsymbol{\varDelta}}(t,\boldsymbol{E}(\cdot))$ which is defined to be the difference

$$\tilde{\boldsymbol{\varDelta}}(t,\boldsymbol{E}(\cdot))=\boldsymbol{\varDelta}(t,\boldsymbol{E}(\cdot))-\boldsymbol{\varDelta}^*(t,\boldsymbol{E}(\cdot))\,. \tag{4.2.52}$$

If follows from the definition (4.2.12) of an isotropic relaxation function and from equation (4.2.50) and the definition (4.2.7) of $\boldsymbol{\varDelta}(t,\boldsymbol{E}(\cdot))$ that

$$\dot{\tilde{\boldsymbol{\varDelta}}}(t,\boldsymbol{E}(\cdot))=(\lambda(0)-\lambda(+\infty))(\operatorname{trace}\boldsymbol{E}(t))\boldsymbol{I}+2(\mu(0)-\mu(+\infty))\dot{\boldsymbol{E}}(t)-k\tilde{\boldsymbol{\varDelta}}(t,\boldsymbol{E})\,. \tag{4.2.53}$$

The conditions (4.2.51) ensure that this relation can be solved uniquely to express $\dot{\boldsymbol{E}}(t)$ as

$$\begin{aligned}\dot{\boldsymbol{E}}(t) &= \frac{1}{2(\mu(0)-\mu(+\infty))}(\dot{\tilde{\boldsymbol{\varDelta}}}(t,\boldsymbol{E}(\cdot))+k\tilde{\boldsymbol{\varDelta}}(t,\boldsymbol{E}(\cdot))\\ &\quad -\nu\big[\operatorname{trace}\dot{\tilde{\boldsymbol{\varDelta}}}(t,\boldsymbol{E}(\cdot))+k\operatorname{trace}\tilde{\boldsymbol{\varDelta}}(t,\boldsymbol{E}(\cdot))\big]\boldsymbol{I}\end{aligned} \tag{4.2.54}$$

where

$$\nu = \frac{\lambda(0)-\lambda(+\infty)}{3(\lambda(0)-\lambda(+\infty))+2(\mu(0)-\mu(+\infty))} \tag{4.2.55}$$

If $\boldsymbol{E}(\cdot)$ is any closed path with $\boldsymbol{E}(t)\equiv\boldsymbol{E}(-\infty)$ for $t\leqslant t_0$ and $\boldsymbol{E}(t_1)=\boldsymbol{E}(-\infty)$ for some $t_1>t_0$ then

$$\begin{aligned}\int_{t_0}^{t_1} \boldsymbol{\varDelta}(t,\boldsymbol{E}(\cdot))\cdot\dot{\boldsymbol{E}}(t)\,dt &= \int_{t_0}^{t_1} (\boldsymbol{\varDelta}^*(\boldsymbol{E}(t))+\tilde{\boldsymbol{\varDelta}}(t,\boldsymbol{E}(\cdot)))\cdot\dot{\boldsymbol{E}}(t)\,dt\\ &= \int_{t_0}^{t_1} \tilde{\boldsymbol{\varDelta}}(t,\boldsymbol{E}(\cdot))\cdot\dot{\boldsymbol{E}}(t)\,dt\end{aligned}$$

and if we substitute the expression (4.2.54) for $\dot{\boldsymbol{E}}(t)$ in to the right-hand side we find that

$$\int_{t_0}^{t_1} \boldsymbol{\Delta}(t,\boldsymbol{E}(\cdot))\cdot\dot{\boldsymbol{E}}(t)\,dt$$

$$= \frac{1}{2(\mu(0)-\mu(+\infty))}\int_{t_0}^{t_1}\Big[\operatorname{trace}\tilde{\boldsymbol{\Delta}}(t,\boldsymbol{E}(\cdot))\dot{\tilde{\boldsymbol{\Delta}}}(t,\boldsymbol{E}(\cdot)) + k\operatorname{trace}(\tilde{\boldsymbol{\Delta}}(t,\boldsymbol{E}(\cdot)))^2$$

$$-\nu\operatorname{trace}\tilde{\boldsymbol{\Delta}}(t,\boldsymbol{E}(\cdot))\operatorname{trace}\dot{\tilde{\boldsymbol{\Delta}}}(t,\boldsymbol{E}(\cdot)) - \nu k(\operatorname{trace}\tilde{\boldsymbol{\Delta}}(t,\boldsymbol{E}(\cdot)))^2\Big]dt\,.$$

Since $\boldsymbol{E}(t)$ is constant for all times $t \leqslant t_0$

$$\tilde{\boldsymbol{\Delta}}(t_0,\boldsymbol{E}(\cdot)) = \boldsymbol{O}$$

and so

$$\int_{t_0}^{t_1} \boldsymbol{\Delta}(t,\boldsymbol{E}(\cdot))\cdot\dot{\boldsymbol{E}}(t)\,dt$$

$$= \frac{1}{4(\mu(0)-\mu(+\infty))}\Big[\operatorname{trace}(\tilde{\boldsymbol{\Delta}}(t_1,\boldsymbol{E}(\cdot)))^2 - \nu(\operatorname{trace}\tilde{\boldsymbol{\Delta}}(t_1,\boldsymbol{E}(\cdot)))^2\Big]$$

$$+ \frac{k}{2(\mu(0)-\mu(+\infty))}\int_{t_0}^{t_1}\Big[\operatorname{trace}(\tilde{\boldsymbol{\Delta}}(t,\boldsymbol{E}(\cdot)))^2 - \nu(\operatorname{trace}\tilde{\boldsymbol{\Delta}}(t,\boldsymbol{E}(\cdot)))^2\Big]dt. \tag{4.2.56}$$

However, if $\lambda(0) > \lambda(+\infty)$ and $\mu(0) > \mu(+\infty)$ the constant ν lies in $0 < \nu < \frac{1}{3}$ and, by the Schwarz inequality,

$$\operatorname{trace}(\tilde{\boldsymbol{\Delta}}(t,\boldsymbol{E}(\cdot)))^2 \geqslant \tfrac{1}{3}(\operatorname{trace}\tilde{\boldsymbol{\Delta}}(t,\boldsymbol{E}(\cdot)))^2 \geqslant \nu(\operatorname{trace}\tilde{\boldsymbol{\Delta}}(t,\boldsymbol{E}(\cdot)))^2$$

which together with (4.2.56) gurantees that the inequality (4.2.16) does hold.

Having verified that all our assumptions hold we turn to evaluating the recoverable entropy. Given any path $\boldsymbol{E}(\cdot)$ for $t \leqslant \tau$ it is easily checked that the integral equation (4.2.31) which determines the reversible extension $\boldsymbol{E}_{\text{rev}}(\cdot)$ has, in this case, the constant solution

$$\boldsymbol{E}_{\text{rev}}(t) \equiv k\int_{-\infty}^{\tau}(\exp k(s-\tau))\,\boldsymbol{E}(s)\,ds \tag{4.2.57}$$

for every $t > \tau$. Since, in general,

$$\boldsymbol{E}(\tau) \neq k\int_{-\infty}^{\tau}(\exp k(s-\tau))\,\boldsymbol{E}(s)\,ds$$

the pointwise limit (4.2.49) usually has a jump discontinuity at $t = \tau$.

On substituting into (4.2.47) the recoverable entropy is found to be

$$\underset{-\infty}{\overset{\tau}{H}}(\boldsymbol{E}(\cdot),e(\cdot))$$
$$=\tfrac{1}{2}(\lambda(0)-\lambda(+\infty))\left[\operatorname{trace}\left(\boldsymbol{E}(\tau)-k\int_{-\infty}^{\tau}(\exp k(s-\tau))\boldsymbol{E}(s)\,ds\right)\right]^2$$
$$+(\mu(0)-\mu(+\infty))\operatorname{trace}\left[\boldsymbol{E}(\tau)-k\int_{-\infty}^{\tau}(\exp k(s-\tau))\boldsymbol{E}(s)\,ds\right]^2 \qquad (4.2.58)$$

which has, as it should, the property that it is non-negative and vanishes if $\boldsymbol{E}(t)$ is constant at all times prior to τ. To within an arbitrary additive constant the entropy is

$$\underset{-\infty}{\overset{\tau}{\eta}}(\boldsymbol{E}(\cdot),e(\cdot))$$
$$=\Theta(e(\tau))-\tfrac{1}{2}\lambda(+\infty)[\operatorname{trace}\boldsymbol{E}(\tau)]^2-\mu(+\infty)\operatorname{trace}[\boldsymbol{E}(\tau)]^2$$
$$-\tfrac{1}{2}(\lambda(0)-\lambda(+\infty))\left[\operatorname{trace}\left(\boldsymbol{E}(\tau)-k\int_{-\infty}^{\tau}(\exp k(s-\tau))\boldsymbol{E}(s)\,ds\right)\right]^2$$
$$-(\mu(0)-\mu(+\infty))\operatorname{trace}\left[\boldsymbol{E}(\tau)-k\int_{-\infty}^{\tau}(\exp k(s-\tau))\boldsymbol{E}(s)\,ds\right]^2 \qquad (4.2.59)$$

In general the entropy recoverable from a path does not vanish, as it does for a material of the differential type, and if the strain and internal energy are held fixed the entropy need not remain constant; it can increase.

The trend functional can also be evaluated explicitly for this material and a short computation shows that

$$\xi(\tau)=k\int_{-\infty}^{\tau}\left\{(\lambda(0)-\lambda(+\infty))\left[\operatorname{trace}\left(\boldsymbol{E}(t)-k\int_{-\infty}^{t}(\exp k(s-t))\boldsymbol{E}(s)\,ds\right)\right]^2\right.$$
$$\left.+2(\mu(0)-\mu(+\infty))\operatorname{trace}\left[\boldsymbol{E}(t)-k\int_{-\infty}^{t}(\exp k(s-t))\boldsymbol{E}(s)\,ds\right]^2\right\}dt \qquad (4.2.60)$$

from which the increasing property of $\xi(\cdot)$ is apparent. An alternative way of writing this result is

$$\xi(\tau)=2k\int_{-\infty}^{\tau}(\eta^*(t)-\eta(t))\,dt \qquad (4.2.61)$$

where $\eta^*(t)=\eta^*(\boldsymbol{E}(t),e(t))$ is the equilibrium entropy at time t. Because, as we have seen, $\eta^*(t)\geqslant\eta(t)$ the increasing property of $\xi(\cdot)$ is also apparent from this formula.

CHAPTER 5

Thermodynamics based on the Clausius-Duhem Inequality

Thus far we have concentrated attention on a theory of thermodynamics which sets out from the thermodynamic inequality (2.2.3) and constructs the entropy explicitly. We turn now to describing an approach to thermodynamics which introduces entropy from the outset and uses the Clausius-Duhem inequality to reduce constitutive equations. This approach was first used by Coleman and Noll in a paper [27] which marks the beginning of the recent revival of activity in continuum thermodynamics. The importance of that paper does not lie in the novelty of its results; indeed for the class of elastic materials with linear viscosity considered by Coleman and Noll all the restrictions derived by them had been either derived before from different starting points or had been conjectured previously. Its importance is that it shows us how to exploit in a systematic way the idea that the second law of thermodynamics requires the Clausius-Duhem inequality to be satisfied in every process which is compatible with the balance laws for mass, momentum, moment of momentum and energy. In principle it will be necessary to supply non-zero body forces and radiation fields to maintain the processes envisaged. If we accept this idea we must test any constitutive equations which claim to describe the behaviour of a material by asking if they are compatible with the Clausius-Duhem inequality; if they are not they must be abandoned. One of the tasks of continuum thermodynamics is to find those materials, within a given class of materials, which are in some sense compatible with thermodynamics. Once we commit ourselves to the viewpoint of Coleman and Noll that compatibility means compatibility with the Clausius-Duhem inequality this task becomes, in a sense, a fairly routine one presenting clearly defined mathematical problems for solution.

The 1963 paper of Coleman and Noll initiated much research into the restrictions imposed by thermodynamics on different kinds of materials. In the same year Coleman and Mizel [18] investigated heat conduction in rigid bodies and Koh and Eringen [54] investigated thermo-viscoelasticity, although their work was not entirely in the spirit of Coleman and Noll's proposal for they made certain thermodynamic assumptions in addition to the Clausius-Duhem inequality. In the following year Coleman and Mizel [19] investigated materials of the rate type, establishing the existence of caloric equations of state for them, and Coleman [8, 9] published work of great importance on materials with fading memory. Coleman's theory was the first rational theory of thermodynamics for materials with memory; the results obtained were subsequently employed in a study of wave propagation[1]. Since 1964 the ideas of Coleman and Noll have been applied by Green and Naghdi [44] to elastic-plastic materials, by Wang and Bowen [72] to materials with quasi-elastic response, by Coleman and Gurtin [15, 16] to materials with internal state variables, by Green and Laws [42], by Laws [56], by Coleman and Mizel [21] to more extensive classes of materials with fading memory, by Gurtin [48], by Owen [61, 62] to materials with an elastic range and by Coleman and Owen [28] to a very large class of materials which have memories but whose memories need not fade with the passage of time.

Many authors have applied the Clausius-Duhem inequality to, for example, theories of mixtures and of liquid crystals but we shall not attempt to cite any of the literature on this subject.

Gurtin [46, 47] has used the Clausius-Duhem inequality in a surprising and interesting way in his investigation of the possibility of spatial interaction in elastic materials and Gurtin and Williams [51] have used the inequality as a basis for showing that the complete symmetry groups of many materials must be subgroups of the unimodular group.

5.1 The Clausius-Duhem Inequality

If the scalar field $\eta(\cdot,\cdot)$ is the *entropy* per unit mass of the body and if $\mathscr{P}$ is any part of the body, consisting always of the same particles, then the rate of increase of the entropy of the part $\mathscr{P}$ is

$$\frac{d}{dt}\int_{\mathscr{P}} \rho\eta\, dV.$$

If the contribution to this rate by heat conduction across the boundary $\partial\mathscr{P}$ of the part $\mathscr{P}$ is taken to be

$$-\int_{\partial\mathscr{P}} \frac{1}{\theta}\boldsymbol{q}\cdot\boldsymbol{n}\, dA$$

[1] See Coleman, Gurtin and Herrera [17], Coleman and Gurtin [14] and Coleman, Greenberg and Gurtin [13].

and if the contribution from the arrival of radiation is taken to be

$$\int_{\mathscr{P}} \frac{1}{\theta} \rho r \, dV$$

then *the Clausius-Duhem inequality asserts that the inequality*

$$\frac{d}{dt} \int_{\mathscr{P}} \rho \eta \, dV \geqslant - \int_{\partial \mathscr{P}} \frac{1}{\theta} \boldsymbol{q} \cdot \boldsymbol{n} \, dA + \int_{\mathscr{P}} \frac{1}{\theta} \rho r \, dV \tag{5.1.1}$$

holds in any process undergone by the body and for any part $\mathscr{P}$ *of it.* An application of the divergence theorem and of the balance law for the mass shows that (5.1.1) is equivalent to the *local Clausius-Duhem inequality*

$$\rho \dot{\eta} \geqslant - \operatorname{div} \left(\frac{1}{\theta} \boldsymbol{q} \right) + \frac{1}{\theta} \rho r . \tag{5.1.2}$$

In any process the body force $\boldsymbol{b}(\cdot,\cdot)$ and the heat supply $r(\cdot,\cdot)$ needed to support the process can be calculated from equations (1.3.14) and (1.3.15), respectively, and indeed, if we substitute for $r(\cdot,\cdot)$ from (1.3.15) into (5.1.2) we obtain the inequality

$$\rho \theta \dot{\eta} \geqslant \frac{1}{\theta} \boldsymbol{g} \cdot \boldsymbol{q} + \rho \dot{e} - \boldsymbol{T} \cdot \boldsymbol{D} \tag{5.1.3}$$

in which $\boldsymbol{D}$ is the rate of strain tensor. In terms of the Helmholtz *free energy* per unit mass, which is

$$\psi = e - \theta \eta , \tag{5.1.4}$$

this inequality assumes the form

$$\dot{\psi} + \frac{1}{\rho \theta} \boldsymbol{g} \cdot \boldsymbol{q} \leqslant \frac{1}{\rho} \boldsymbol{T} \cdot \boldsymbol{D} - \eta \dot{\theta} \tag{5.1.5}$$

or, in terms of the Piola-Kirchhoff stress tensor $\boldsymbol{S}$ introduced in (2.2.5), the form

$$\dot{\psi} + \frac{1}{\rho \theta} \boldsymbol{g} \cdot \boldsymbol{q} \leqslant \boldsymbol{S} \cdot \dot{\boldsymbol{F}} - \eta \dot{\theta} . \tag{5.1.6}$$

The Clausius-Duhem inequality (5.1.1) with the term

$$\int_{\mathscr{P}} \frac{1}{\theta} \rho r \, dV$$

appearing is due to Truesdell and Toupin [69]. The techniques which determine the restrictions which the inequality imposes on constitutive equations are not entirely special to this inequality and various proposals have been made which generalise the inequality.

One such generalisation is that of Green and Laws [43] who have investigated the restrictions imposed on thermoelastic materials and Newtonian fluids by assuming the global Clausius-Duhem inequality (5.1.1) for the whole body only and not for its parts. In their study of the axiomatic foundations of continuum thermodynamics Gurtin and Williams [53][1] were led to distinguish two temperatures in a material, a radiative temperature θ and a conductive temperature φ and they were led to the inequality

$$\rho\dot{\eta} \geqslant -\operatorname{div}\left(\frac{1}{\varphi}\boldsymbol{q}\right)+\frac{1}{\theta}\rho r$$

rather than to (5.1.2). However, as Gurtin and Williams showed in [52], if the material is simple it generally happens that the two temperatures coincide, whereas they need not coincide if it is not simple[2].

At about the same time Müller [60] suggested that the term

$$-\int_{\partial\mathscr{P}} \frac{1}{\theta}\boldsymbol{q}\cdot\boldsymbol{n}\,dA$$

which we have taken to be the contribution to the rate of increase of entropy in the part $\mathscr{P}$ due to heat conduction across the boundary be replaced by a more general term

$$-\int_{\partial\mathscr{P}} \boldsymbol{j}\cdot\boldsymbol{n}\,dA\,,$$

where $\boldsymbol{j}$ is the entropy flux vector and which is to be determined by an additional constitutive equation. If the material is simple and has fading memory it turns out, unless it has unusual symmetry properties, that $\boldsymbol{j}=(1/\theta)\boldsymbol{q}$ but for non-simple materials $\boldsymbol{j}$ does not necessarily reduce to $(1/\theta)\boldsymbol{q}$.

Yet another starting point for a theory of thermodynamics is provided by the inequality of Meixner [58] and called by him the fundamental inequality. Meixner is unwilling to concede the existence of entropy away from thermodynamic equilibrium and his inequality involves only the equilibrium entropy. To date not many consequences of this inequality have been elaborated.

We shall study simple materials in which the stress tensor $\boldsymbol{T}$, the heat flux vector $\boldsymbol{q}$, the internal energy e and the entropy η are determined

---

[1] See also Williams [73, 74] and Fisher and Leitmann [41].

[2] See Chen and Gurtin [4] and Chen, Gurtin and Williams [5, 6].

by the histories of the deformation gradient $\boldsymbol{F}$, the absolute temperature θ and the spatial temperature gradient $\boldsymbol{g}$ through constitutive equations of the kind already considered in section 1.4, namely

$$\boldsymbol{T}(t)= \underset{s=-\infty}{\overset{t}{\boldsymbol{T}}}(\boldsymbol{F}(s),\theta(s),\boldsymbol{g}(s))= \underset{-\infty}{\overset{t}{\boldsymbol{T}}}(\boldsymbol{F}(\cdot),\theta(\cdot),\boldsymbol{g}(\cdot)), \tag{5.1.7}$$

$$\boldsymbol{q}(t)= \underset{s=-\infty}{\overset{t}{\boldsymbol{q}}}(\boldsymbol{F}(s),\theta(s),\boldsymbol{g}(s))= \underset{-\infty}{\overset{t}{\boldsymbol{q}}}(\boldsymbol{F}(\cdot),\theta(\cdot),\boldsymbol{g}(\cdot)), \tag{5.1.8}$$

$$e(t)= \underset{s=-\infty}{\overset{t}{e}}(\boldsymbol{F}(s),\theta(s),\boldsymbol{g}(s))= \underset{-\infty}{\overset{t}{e}}(\boldsymbol{F}(\cdot),\theta(\cdot),\boldsymbol{g}(\cdot)), \tag{5.1.9}$$

$$\eta(t)= \underset{s=-\infty}{\overset{t}{\eta}}(\boldsymbol{F}(s),\theta(s),\boldsymbol{g}(s))= \underset{-\infty}{\overset{t}{\eta}}(\boldsymbol{F}(\cdot),\theta(\cdot),\boldsymbol{g}(\cdot)). \tag{5.1.10}$$

Necessarily the free energy is given by a relation

$$\psi(t)= \underset{s=-\infty}{\overset{t}{\psi}}(\boldsymbol{F}(s),\theta(s),\boldsymbol{g}(s))= \underset{-\infty}{\overset{t}{\psi}}(\boldsymbol{F}(\cdot),\theta(\cdot),\boldsymbol{g}(\cdot)) \tag{5.1.11}$$

in which the response functional $\underset{-\infty}{\overset{t}{\psi}}$ is determined by the functionals $\underset{-\infty}{\overset{t}{e}}$ and $\underset{-\infty}{\overset{t}{\eta}}$ by way of definition (5.1.4).

The requirement that the inequality (5.1.6) must hold in any process severely restricts the behaviour of the response functionals for the stress, the heat flux, the internal energy, the entropy and the free energy and we shall illustrate how these restrictions can be found by studying two classes of materials, materials of the differential type and complexity 1 and materials with long-range memory which respond elastically to sudden changes in the values of $\boldsymbol{F}$, θ and $\boldsymbol{g}$.

5.2 Materials of the Differential Type

For materials of the differential type and complexity 1 equations (5.1.7), (5.1.8), (5.1.9), (5.1.10) and (5.1.11) assume the forms

$$\boldsymbol{T}(t)=\boldsymbol{T}(\boldsymbol{F}(t),\dot{\boldsymbol{F}}(t),\theta(t),\boldsymbol{g}(t)), \tag{5.2.1}$$

$$\boldsymbol{q}(t)=\boldsymbol{q}(\boldsymbol{F}(t),\dot{\boldsymbol{F}}(t),\theta(t),\boldsymbol{g}(t)), \tag{5.2.2}$$

$$e(t)=e(\boldsymbol{F}(t),\dot{\boldsymbol{F}}(t),\theta(t),\boldsymbol{g}(t)), \tag{5.2.3}$$

$$\eta(t)=\eta(\boldsymbol{F}(t),\dot{\boldsymbol{F}}(t),\theta(t),\boldsymbol{g}(t)), \tag{5.2.4}$$

and

$$\psi(t)=\psi(\boldsymbol{F}(t),\dot{\boldsymbol{F}}(t),\theta(t),\boldsymbol{g}(t)). \tag{5.2.5}$$

Both thermoelastic materials, for which there is no dependence upon the rate $\dot{\boldsymbol{F}}$, and linearly viscous fluids are materials of this kind. For the linearly viscous fluid the constitutive relations reduce to the relations (1.4.29), (1.4.30), (1.4.31) and (1.4.32) together with

$$\psi = \psi(v, \theta)\,.$$

It is convenient to work not with the stress tensor $\boldsymbol{T}$ but with the Piola-Kirchhoff stress tensor for which

$$\boldsymbol{S}(t) = \boldsymbol{S}(\boldsymbol{F}(t), \dot{\boldsymbol{F}}(t), \boldsymbol{g}(t))\,. \tag{5.2.6}$$

If we differentiate both sides of (5.2.5) with respect to the time t and use the chain-rule we find that

$$\dot{\psi} = \frac{\partial \psi}{\partial \boldsymbol{F}} \cdot \dot{\boldsymbol{F}} + \frac{\partial \psi}{\partial \dot{\boldsymbol{F}}} \cdot \ddot{\boldsymbol{F}} + \frac{\partial \psi}{\partial \theta}\dot{\theta} + \frac{\partial \psi}{\partial \boldsymbol{g}} \cdot \dot{\boldsymbol{g}} \tag{5.2.7}$$

or, in the suffix notation, that

$$\dot{\psi} = \sum_{i,j} \left(\frac{\partial \psi}{\partial F_{ij}} \dot{F}_{ij} + \frac{\partial \psi}{\partial \dot{F}_{ij}} \ddot{F}_{ij} \right) + \frac{\partial \psi}{\partial \theta}\dot{\theta} + \sum_i \frac{\partial \psi}{\partial g_i} \dot{g}_i$$

and on substituting from (5.2.7) into (5.1.6) we obtain the inequality

$$\left(\frac{\partial \psi}{\partial \boldsymbol{F}} - \boldsymbol{S} \right) \cdot \dot{\boldsymbol{F}} + \frac{\partial \psi}{\partial \dot{\boldsymbol{F}}} \cdot \ddot{\boldsymbol{F}} + \left(\frac{\partial \psi}{\partial \theta} + \eta \right) \dot{\theta} + \frac{\partial \psi}{\partial \boldsymbol{g}} \cdot \dot{\boldsymbol{g}} + \frac{1}{\rho \theta} \boldsymbol{g} \cdot \boldsymbol{q} \leqslant 0\,. \tag{5.2.8}$$

However, we have seen already in section 4.1 that if $\boldsymbol{M}_0$ is any tensor with $\det \boldsymbol{M}_0 > 0$, if $\boldsymbol{M}_1$ and $\boldsymbol{M}_2$ are any tensors, if $\alpha_0 > 0$ is any positive scalar, if α_1 is any scalar and if $\boldsymbol{w}_0$ and $\boldsymbol{w}_1$ are any vectors and t_0 is any time then there is a process in which

$$\begin{aligned} &\boldsymbol{F}(t_0) = \boldsymbol{M}_0, \quad &&\dot{\boldsymbol{F}}(t_0) = \boldsymbol{M}_1, \quad &&\ddot{\boldsymbol{F}}(t_0) = \boldsymbol{M}_2, \\ &\theta(t_0) = \alpha_0, &&\dot{\theta}(t_0) = \alpha_1, \\ &\boldsymbol{g}(t_0) = \boldsymbol{w}_0, &&\dot{\boldsymbol{g}}(t_0) = \boldsymbol{w}_1, \end{aligned}$$

and so the inequality (5.2.8) states that

$$\begin{aligned} &\left(\frac{\partial \psi}{\partial \boldsymbol{F}}(\boldsymbol{M}_0, \boldsymbol{M}_1, \alpha_0, \boldsymbol{w}_0) - \boldsymbol{S}(\boldsymbol{M}_0, \boldsymbol{M}_1, \alpha_0, \boldsymbol{w}_0) \right) \cdot \boldsymbol{M}_1 \\ &\qquad + \frac{\partial \psi}{\partial \dot{\boldsymbol{F}}}(\boldsymbol{M}_0, \boldsymbol{M}_1, \alpha_0, \boldsymbol{w}_0) \cdot \boldsymbol{M}_2 \\ &\qquad + \left(\frac{\partial \psi}{\partial \theta}(\boldsymbol{M}_0, \boldsymbol{M}_1, \alpha_0, \boldsymbol{w}_0) + \eta(\boldsymbol{M}_0, \boldsymbol{M}_1, \alpha_0, \boldsymbol{w}_0) \right) \alpha_1 \\ &\qquad + \frac{\partial \psi}{\partial \boldsymbol{g}}(\boldsymbol{M}_0, \boldsymbol{M}_1, \alpha_0, \boldsymbol{w}_0) \cdot \boldsymbol{w}_1 \\ &\qquad + \frac{1}{\alpha_0} \boldsymbol{w}_0 \cdot \boldsymbol{q}(\boldsymbol{M}_0, \boldsymbol{M}_1, \alpha_0, \boldsymbol{w}_0) \leqslant 0\,. \end{aligned} \tag{5.2.9}$$

This inequality can hold for all choices of the tensor $\boldsymbol{M}_2$ only if

$$\frac{\partial\psi}{\partial\dot{\boldsymbol{F}}}\equiv\boldsymbol{O} \tag{5.2.10}$$

and for all choices of the vector $\boldsymbol{w}_1$ only if

$$\frac{\partial\psi}{\partial\boldsymbol{g}}\equiv\boldsymbol{0}. \tag{5.2.11}$$

In other words *the free energy must be independent of the rate* $\dot{\boldsymbol{F}}$ *and of the temperature gradient* and the inequality (5.2.9) reduces to the inequality

$$\begin{aligned}\left(\frac{\partial\psi}{\partial\boldsymbol{F}}(\boldsymbol{M}_0,\alpha_0)-\boldsymbol{S}(\boldsymbol{M}_0,\boldsymbol{M}_1,\alpha_0,\boldsymbol{w}_0)\right)\cdot\boldsymbol{M}_1&\\ +\left(\frac{\partial\psi}{\partial\theta}(\boldsymbol{M}_0,\alpha_0)+\eta(\boldsymbol{M}_0,\boldsymbol{M}_1,\alpha_0,\boldsymbol{w}_0)\right)\alpha_1&\\ +\frac{1}{\alpha_0}\boldsymbol{w}_0\cdot\boldsymbol{q}(\boldsymbol{M}_0,\boldsymbol{M}_1,\alpha_0,\boldsymbol{w}_0)\leqslant 0&\end{aligned}$$

which can hold for every choice of the scalar α_1 only if *the entropy is determined by the free energy by way of the relation*

$$\eta(\boldsymbol{F},\theta)=-\frac{\partial\psi}{\partial\theta}(\boldsymbol{F},\theta). \tag{5.2.12}$$

Consequently the entropy has to be independent of the rate $\dot{\boldsymbol{F}}$ *and of the temperature gradient* $\boldsymbol{g}$. Equation (5.2.9) reduces to the inequality

$$\left(\frac{\partial\psi}{\partial\boldsymbol{F}}(\boldsymbol{M}_0,\alpha_0)-\boldsymbol{S}(\boldsymbol{M}_0,\boldsymbol{M}_1,\alpha_0,\boldsymbol{w}_0)\right)\cdot\boldsymbol{M}_1+\frac{1}{\alpha_0}\boldsymbol{w}_0\cdot\boldsymbol{q}(\boldsymbol{M}_0,\boldsymbol{M}_1,\alpha_0,\boldsymbol{w}_0)\leqslant 0. \tag{5.2.13}$$

On setting $\boldsymbol{w}_0=\boldsymbol{0}$ in (5.2.13), replacing M_1 by λM_1, where λ is any positive scalar, dividing both sides of the inequality which results by λ and taking the limit as $\lambda\to 0$ we find that

$$\left(\frac{\partial\psi}{\partial\boldsymbol{F}}(\boldsymbol{M}_0,\alpha_0)-\boldsymbol{S}(\boldsymbol{M}_0,\boldsymbol{O},\alpha_0,\boldsymbol{0})\right)\cdot\boldsymbol{M}_1\leqslant 0$$

and so

$$\frac{\partial\psi}{\partial\boldsymbol{F}}(\boldsymbol{M}_0,\alpha_0)=\boldsymbol{S}(\boldsymbol{M}_0,\boldsymbol{O},\alpha_0,\boldsymbol{0}).$$

This means that *the Piola-Kirchhoff stress*

$$\boldsymbol{S}^*(\boldsymbol{F},\theta)=\boldsymbol{S}(\boldsymbol{F},\boldsymbol{O},\theta,\boldsymbol{0}) \tag{5.2.14}$$

which are special cases of the equations (5.1.7), (5.1.8), (5.1.9), (5.1.10) and (5.1.11) in that the dependence upon the temperature gradient is only through its present value and not on all of its history. For the present we shall work with the more general relations (5.1.7), (5.1.8), (5.1.9), (5.1.10) and (5.1.11) and admit dependence upon the history of the temperature gradient.

It is crucial, if the results of this section are to hold, that we restrict our attention to materials for which the stress, the heat flux, the internal energy and the entropy depend continuously on this histories of the deformation gradient, the temperature and the temperature gradient and for which the free energy depends on these histories in a continuously differentiable way. The meaning to be attached to continuity and continuous differentiability will be explained shortly. Because the free energy is continuously differentiable it has a certain chain-rule property; the importance of the chain-rule has been stressed by Wang and Bowen [72] and by Gurtin [48] whose work brings out, in a very clear way, the fact that the Clausius-Duhem inequality and the chain-rule for the free energy are the two essential ingredients in Coleman's approach to the thermodynamics of materials with memory.

For the sake of brevity we introduce the vector space $\mathscr{V}$ of all triples $(\boldsymbol{A}, a, \boldsymbol{a})$ consisting of a tensor $\boldsymbol{A}$, a scalar a and a vector $\boldsymbol{a}$. This space has dimension 13, the *scalar product* of two triples $(\boldsymbol{A}, a, \boldsymbol{a})$ and $(\boldsymbol{B}, b, \boldsymbol{b})$ is

$$(\boldsymbol{A}, a, \boldsymbol{a}) \cdot (\boldsymbol{B}, b, \boldsymbol{b}) = \boldsymbol{A} \cdot \boldsymbol{B} + ab + \boldsymbol{a} \cdot \boldsymbol{b} = \operatorname{trace} \boldsymbol{A} \boldsymbol{B}^T + ab + \boldsymbol{a} \cdot \boldsymbol{b} \quad (5.3.6)$$

and the *norm* of the triple $(\boldsymbol{A}, a, \boldsymbol{a})$ is

$$|(\boldsymbol{A}, a, \boldsymbol{a})| = [(\boldsymbol{A}, a, \boldsymbol{a}) \cdot (\boldsymbol{A}, a, \boldsymbol{a})]^{\frac{1}{2}}. \quad (5.3.7)$$

If, as always, $\boldsymbol{F}$ is the deformation gradient, if θ is the abolute temperature and if $\boldsymbol{g}$ is the temperature gradient then $(\boldsymbol{F}, \theta, \boldsymbol{g})$ certainly lies in $\mathscr{V}$; in fact $(\boldsymbol{F}, \theta, \boldsymbol{g})$ lies in the subset $\mathscr{U}$ of $\mathscr{V}$ consisting of those triples $(\boldsymbol{A}, a, \boldsymbol{a})$ with $\det \boldsymbol{A} > 0$ and $a > 0$; $\mathscr{U}$ is both open and connected.

If we write $\boldsymbol{\Lambda}(\cdot) = (\boldsymbol{F}(\cdot), \theta(\cdot). \boldsymbol{g}(\cdot))$ the constitutive relations lead us to studying functional relations of the form

$$f(t) = \underset{0 < s < +\infty}{f} (\boldsymbol{\Lambda}(t), \boldsymbol{\Lambda}^t(s)) = \bar{f}(\boldsymbol{\Lambda}(t), \boldsymbol{\Lambda}^t(\cdot)) \quad (5.3.8)$$

in which the dependence upon the present value $\boldsymbol{\Lambda}(t) = (\boldsymbol{F}(t), \theta(t), \boldsymbol{g}(t))$ has been made explicit and the function $\boldsymbol{\Lambda}^t(\cdot)$ is the *past history* up to time t, that is to say

$$\boldsymbol{\Lambda}^t(s) = \boldsymbol{\Lambda}(t-s), \qquad s > 0, \quad (5.3.9)$$

and $\boldsymbol{\Lambda}^t(s)$ is the value of $\boldsymbol{\Lambda}(\cdot)$ at a time s units before t. Of course the values of the past history lie in the set $\mathscr{U}$. Our aim, to begin with, is to produce a chain-rule for the functional $\bar{f}$ occurring in (5.3.8) of the kind proved by Mizel and Wang [59] although, for simplicity, we shall make somewhat stronger assumptions than they did and we shall prove a somewhat weaker result than they did.

The idea that the material has a fading memory can be made precise with the aid of an oblivator or influence function[1]. An oblivator is a continuous, positive, monotone decreasing function $\gamma(\cdot)$ with

$$\int_0^{+\infty} \gamma(s)^2\, ds < +\infty .$$

It is to be regarded as characterising the rate at which the memory of the material fades with the passage of time. It turns out that the results we prove are independent of the particular choice of the oblivator which suggests that there are more sophisticated ways of defining fading memory in which the oblivator itself does not play such a prominent role. That is so and the reader is referred to the papers of Wang [70, 71] and of Coleman and Mizel [20, 22], for an account of more general theories of fading memory.

If $\gamma(\cdot)$ is an oblivator the collection of all measurable functions $\boldsymbol{\Gamma}(\cdot)$, defined for every $s>0$, whose values lie in the space $\mathscr{V}$ and for which the norm

$$\|\boldsymbol{\Gamma}(\cdot)\| = \left[\int_0^{+\infty} \gamma(s)^2\, \boldsymbol{\Gamma}(s)\cdot\boldsymbol{\Gamma}(s)\, ds\right]^{\frac{1}{2}} \tag{5.3.10}$$

is finite can be regarded as a Hilbert space $\mathscr{H}$ in which the scalar product of the two functions $\boldsymbol{\Gamma}_1(\cdot)$ and $\boldsymbol{\Gamma}_2(\cdot)$ is

$$\langle \boldsymbol{\Gamma}_1(\cdot), \boldsymbol{\Gamma}_2(\cdot)\rangle = \int_0^{+\infty} \gamma(s)^2\, \boldsymbol{\Gamma}_1(s)\cdot\boldsymbol{\Gamma}_2(s)\, ds . \tag{5.3.11}$$

Because the oblivator is monotone decreasing the norm $\|\boldsymbol{\Gamma}(\cdot)\|$, computed from the formula (5.3.10), attaches greater weight to the values $\boldsymbol{\Gamma}(s)$ taken by $\boldsymbol{\Gamma}(\cdot)$ when s is small than it does to the values taken when s is large. This means that the difference of the functions $\boldsymbol{\Gamma}_1(\cdot)$ and $\boldsymbol{\Gamma}_2(\cdot)$ can have a small norm $\|\boldsymbol{\Gamma}_1(\cdot)-\boldsymbol{\Gamma}_2(\cdot)\|$ only if, loosely speaking, $\boldsymbol{\Gamma}_1(s)$ and $\boldsymbol{\Gamma}_2(s)$ coincide very nearly for small values of s; they need not be close to each other for large values of s. In particular if $\boldsymbol{\Gamma}_1(\cdot)$ and $\boldsymbol{\Gamma}_2(\cdot)$ are past histories, let us say $\boldsymbol{\Gamma}_1(\cdot)=\boldsymbol{\Lambda}_1^t(\cdot)$ and $\boldsymbol{\Gamma}_2(\cdot)=\boldsymbol{\Lambda}_2^t(\cdot)$, then $\boldsymbol{\Lambda}_1(u)$ and $\boldsymbol{\Lambda}_2(u)$ must nearly coincide at times u just prior to t.

We suppose the functional $\bar{f}(\boldsymbol{\Lambda}, \boldsymbol{\Gamma}(\cdot))$ to be defined for every $\boldsymbol{\Lambda}$ in $\mathscr{U}$ and for every function $\boldsymbol{\Gamma}(\cdot)$ in $\mathscr{H}$ which is *suitable* in the sense that $\boldsymbol{\Gamma}(s)$

[1] Coleman and Noll [25] were the first to define fading memory by using an oblivator or influence function, as they called it.

lies in $\mathscr{U}$ for almost every $s>0$. Then $\bar{f}$ is said to be *continuous* if for each fixed and suitable $\boldsymbol{\Gamma}(\cdot)$ and for every $\boldsymbol{\chi}(\cdot)$ such that $\boldsymbol{\Gamma}(\cdot)+\boldsymbol{\chi}(\cdot)$ is suitable

$$\bar{f}(\boldsymbol{\Lambda}+\boldsymbol{\Pi},\boldsymbol{\Gamma}(\cdot)+\boldsymbol{\chi}(\cdot))-f(\boldsymbol{\Lambda},\boldsymbol{\Gamma}(\cdot))\to 0 \tag{5.3.12}$$

as $|\boldsymbol{\Pi}|+\|\boldsymbol{\chi}(\cdot)\|\to 0$. The continuity of $\bar{f}$ ensures that if $\boldsymbol{\Lambda}_1(\cdot)$, $\boldsymbol{\Lambda}_2(\cdot)$ are two functions with values in $\mathscr{U}$, if the past histories $\boldsymbol{\Lambda}_1^t(\cdot)$, $\boldsymbol{\Lambda}_2^t(\cdot)$ lie in $\mathscr{H}$ and if $|\boldsymbol{\Lambda}_1(t)-\boldsymbol{\Lambda}_2(t)|$ and $\|\boldsymbol{\Lambda}_1^t(\cdot)-\boldsymbol{\Lambda}_2^t(\cdot)\|$ are both small enough then $\bar{f}(\boldsymbol{\Lambda}_1(t),\boldsymbol{\Lambda}_1^t(\cdot))$ and $\bar{f}(\boldsymbol{\Lambda}_2(t),\boldsymbol{\Lambda}_2^t(\cdot))$ are close together. The functional $\bar{f}$ is said to be *continuously differentiable* if the relation (5.3.12) can be replaced by the relation

$$\frac{1}{|\boldsymbol{\Pi}|+\|\boldsymbol{\chi}(\cdot)\|}\Big|\bar{f}(\boldsymbol{\Lambda}+\boldsymbol{\Pi},\boldsymbol{\Gamma}(\cdot)+\boldsymbol{\chi}(\cdot))-\bar{f}(\boldsymbol{\Lambda},\boldsymbol{\Gamma}(\cdot)) - D\bar{f}(\boldsymbol{\Lambda},\boldsymbol{\Gamma}(\cdot))\cdot\boldsymbol{\Pi}-\delta\bar{f}(\boldsymbol{\Lambda},\boldsymbol{\Gamma}(\cdot)|\boldsymbol{\chi}(\cdot))\Big|\to 0 \tag{5.3.13}$$

as $|\boldsymbol{\Pi}|+\|\boldsymbol{\chi}(\cdot)\|\to 0$. Here $D\bar{f}(\boldsymbol{\Lambda},\boldsymbol{\Gamma}(\cdot))$ is a continuous functional with values in the space $\mathscr{V}$ and, for each fixed $\boldsymbol{\Lambda}$ and $\boldsymbol{\Gamma}(\cdot)$, $\delta\bar{f}(\boldsymbol{\Lambda},\boldsymbol{\Gamma}(\cdot)|\boldsymbol{\chi}(\cdot))$ is a continuous scalar-valued functional depending linearly on $\boldsymbol{\chi}(\cdot)$ and defined on the closed subspace of $\mathscr{H}$ spanned by the functions $\boldsymbol{\chi}(\cdot)$ for which $\boldsymbol{\Gamma}(\cdot)+\boldsymbol{\chi}(\cdot)$ is suitable. It is assumed too that $\delta\bar{f}(\boldsymbol{\Lambda},\boldsymbol{\Gamma}(\cdot)|\boldsymbol{\chi}(\cdot))$ depends continuously on the pair $(\boldsymbol{\Lambda},\boldsymbol{\Gamma}(\cdot))$. Of course $D\bar{f}(\boldsymbol{\Lambda},\boldsymbol{\Gamma}(\cdot))$ is nothing but the partial derivative of $\bar{f}(\boldsymbol{\Lambda},\boldsymbol{\Gamma}(\cdot))$ taken with respect to $\boldsymbol{\Lambda}$ with $\boldsymbol{\Gamma}(\cdot)$ held fixed, that is to say

$$D\bar{f}(\boldsymbol{\Lambda},\boldsymbol{\Gamma}(\cdot))=\frac{\partial\bar{f}}{\partial\boldsymbol{\Lambda}}(\boldsymbol{\Lambda},\boldsymbol{\Gamma}(\cdot)). \tag{5.3.14}$$

We are now in a position to prove a chain-rule which is weaker than the one proved by Mizel and Wang [59] but which can be proved in a more straighforward way and is adequate for our purposes[1].

Chain-Rule. Suppose that the functional $\bar{f}$ is continuously differentiable and that $\boldsymbol{\Lambda}(\cdot)$ is a function or $(-\infty,+\infty)$ *satisfying the chain-rule conditions, that is its values are in $\mathscr{U}$, it has two continuous derivatives $\dot{\boldsymbol{\Lambda}}(\cdot)$,*

[1] In their study of materials with quasi-elastic response Wang and Bowen [72] avoided proving a chain-rule by the ingenious device of assuming from the outset that for each fixed $\boldsymbol{\Lambda}(\cdot)$ in some unspecified class of functions the function

$$f(\boldsymbol{\Omega},t)=\bar{f}(\boldsymbol{\Omega},\boldsymbol{\Lambda}^t(\cdot))$$

is a continuously differentiable function of $\boldsymbol{\Omega}$ and t. Trivially then

$$\dot{f}(t)=\frac{d}{dt}f(\boldsymbol{\Lambda}(t),t)=\frac{\partial f}{\partial\boldsymbol{\Lambda}}(\boldsymbol{\Lambda}(t),t)+\frac{\partial f}{\partial t}(\boldsymbol{\Lambda}(t),t).$$

A chain-rule was also adopted as an axiom by Gurtin [48].

$\ddot{\Lambda}(\cdot)$ *and for every t the past histories*[1] $\Lambda^t(\cdot)$, $\dot{\Lambda}^t(\cdot)$, $\ddot{\Lambda}^t(\cdot)$ *are in* $\mathscr{H}$. *Then the function* $f(t)=\bar{f}(\Lambda(t),\Lambda^t(\cdot))$ *is continuously differentiable and its derivative is*

$$\dot{f}(t)=D\bar{f}(\Lambda(t),\Lambda^t(\cdot))\cdot\dot{\Lambda}(t)+\delta\bar{f}(\Lambda(t),\Lambda^t(\cdot))\,|\,\dot{\Lambda}^t(\cdot)). \tag{5.3.15}$$

Observe to begin with that for each fixed t the past history $\Lambda^t(\cdot)$ is certainly suitable if the hypotheses on $\Lambda(\cdot)$ are fulfilled and so

$$f(t)=\bar{f}(\Lambda(t),\Lambda^t(\cdot))$$

is well-defined. We shall show below that

$$\|\Lambda^{t+\alpha}(\cdot)-\Lambda^t(\cdot)\|\to 0 \tag{5.3.16}$$

as $\alpha\to 0$; accordingly if

$$\begin{aligned}\mathscr{R}(\alpha)&=f(t+\alpha)-f(t)-D\bar{f}(\Lambda(t),\Lambda^t(\cdot))\cdot(\Lambda(t+\alpha)-\Lambda(t))\\&\quad-\delta\bar{f}(\Lambda(t),\Lambda^t(\cdot)\,|\,\Lambda^{t+\alpha}(\cdot)-\Lambda^t(\cdot))\\&=\bar{f}(\Lambda(t+\alpha),\Lambda^{t+\alpha}(\cdot))-\bar{f}(\Lambda(t),\Lambda^t(\cdot))\\&\quad-D\bar{f}(\Lambda(t),\Lambda^t(\cdot))\cdot(\Lambda(t+\alpha)-\Lambda(t))\\&\quad-\delta\bar{f}(\Lambda(t),\Lambda^t(\cdot)\,|\,\Lambda^{t+\alpha}(\cdot)-\Lambda^t(\cdot))\end{aligned} \tag{5.3.17}$$

then

$$\frac{\mathscr{R}(\alpha)}{|\Lambda(t+\alpha)-\Lambda(t)|+\|\Lambda^{t+\alpha}(\cdot)-\Lambda^t(\cdot)\|}\to 0 \tag{5.3.18}$$

as $\alpha\to 0$. However

$$\begin{aligned}\frac{\mathscr{R}(\alpha)}{\alpha}&=\frac{\mathscr{R}(\alpha)}{|\Lambda(t+\alpha)-\Lambda(t)|+\|\Lambda^{t+\alpha}(\cdot)-\Lambda^t(\cdot)\|}\\&\quad\times\frac{1}{\alpha}\{|\Lambda(t+\alpha)-\Lambda(t)|+\|\Lambda^{t+\alpha}(\cdot)-\Lambda^t\|\}\end{aligned} \tag{5.3.19}$$

and if we can show, in addition to (5.3.16), that

$$\left\|\frac{1}{\alpha}(\Lambda^{t+\alpha}(\cdot)-\Lambda^t(\cdot))-\dot{\Lambda}^t(\cdot)\right\|\to 0 \tag{5.3.20}$$

as $\alpha\to 0$ it follows first of all that

$$\frac{1}{\alpha}\{|\Lambda(t+\alpha)-\Lambda(t)|+\|\Lambda^{t+\alpha}(\cdot)-\Lambda^t(\cdot)\|\}\to|\dot{\Lambda}(t)|+\|\dot{\Lambda}^t(\cdot)\|$$

---

[1] The symbols $\dot{\Lambda}^t(\cdot)$ and $\ddot{\Lambda}^t(\cdot)$ stand for the past histories of the derivatives $\dot{\Lambda}(\cdot)$ and $\ddot{\Lambda}(\cdot)$ so that $\dot{\Lambda}^t(s)=\dot{\Lambda}(t-s)$, $\ddot{\Lambda}^t(s)=\ddot{\Lambda}(t-s)$ $(s>0)$; thus $\dot{\Lambda}^t(\cdot)$ is not to be confused with the derivative of the past history $\Lambda^t(\cdot)$ which is

$$\frac{d}{ds}\Lambda^t(s)=\frac{d}{ds}\Lambda(t-s)=-\dot{\Lambda}(t-s)=-\dot{\Lambda}^t(s).$$

as $\alpha \to 0$ and hence from (5.3.17) and (5.3.18) that

$$\frac{\mathcal{R}(\alpha)}{\alpha} \to 0 \tag{5.3.21}$$

as $\alpha \to 0$ and secondly because of the assumptions made on $\bar{f}$ that

$$\delta \bar{f}(\Lambda(t), \Lambda^t(\cdot) \,|\, \frac{1}{\alpha}(\Lambda^{t+\alpha}(\cdot) - \Lambda^t(\cdot))) \to \delta \bar{f}(\Lambda(t), \Lambda^t(\cdot) \,|\, \dot{\Lambda}^t(\cdot)) \tag{5.3.22}$$

as $\alpha \to 0$. If we now divide both sides of (5.3.17) by α, take the limit as $\alpha \to 0$ and use (5.3.21) and (5.3.22) we obtain the required formula (5.3.15) for the derivative of $f(\cdot)$.

It remains to be shown that (5.3.16) and (5.3.20) do hold. We have

$$\begin{aligned}\|\Lambda^{t+\alpha}(\cdot) - \Lambda^t(\cdot)\|^2 &= \int_0^{+\infty} \gamma(s)^2 |\Lambda(t+\alpha-s) - \Lambda(t-s)|^2 \, ds \\ &= \int_0^{+\infty} \gamma(s)^2 \left| \int_0^{\alpha} \dot{\Lambda}(t-s+u)\, du \right|^2 ds. \end{aligned} \tag{5.3.23}$$

But, by the Schwarz inequality,

$$\left| \int_0^{\alpha} \dot{\Lambda}(t-s+u)\, du \right|^2 \leqslant \alpha \int_0^{\alpha} |\dot{\Lambda}(t-s+u)|^2 \, du$$

and so on changing the order of integration in (5.3.23) we find that

$$\begin{aligned}\|\Lambda^{t+\alpha}(\cdot) - \Lambda^t(\cdot)\|^2 &\leqslant \alpha \int_0^{\alpha} \left\{ \int_0^{+\infty} \gamma(s)^2 |\dot{\Lambda}^{t+u}(s)|^2 \, ds \right\} du \\ &= \alpha \int_0^{\alpha} \|\dot{\Lambda}^{t+u}(\cdot)\|^2 \, du, \end{aligned} \tag{5.3.24}$$

which proves (5.3.16).

To prove (5.3.20) note that

$$\begin{aligned}&\left\| \frac{1}{\alpha}(\Lambda^{t+\alpha}(\cdot) - \Lambda^t(\cdot)) - \dot{\Lambda}^t(\cdot) \right\|^2 \\ &\qquad = \int_0^{+\infty} \gamma(s)^2 \left| \frac{1}{\alpha}(\Lambda(t+\alpha-s) - \Lambda(t-s)) - \dot{\Lambda}(t-s) \right|^2 ds \\ &\qquad = \int_0^{+\infty} \gamma(s)^2 \left| \frac{1}{\alpha} \int_0^{\alpha} (\dot{\Lambda}(t-s+u) - \dot{\Lambda}(t-s))\, du \right|^2 ds. \end{aligned} \tag{5.3.25}$$

By the Schwarz inequality

$$\left|\int_0^\alpha (\dot{\Lambda}(t-s+u)-\dot{\Lambda}(t-s))\,du\right|^2 \leqslant \alpha\int_0^\alpha |\dot{\Lambda}(t-s+u)-\dot{\Lambda}(t-s)|^2\,du$$

and on interchanging the orders of integration in (5.3.25) we find that

$$\left\|\frac{1}{\alpha}(\Lambda^{t+\alpha}(\cdot)-\Lambda^t(\cdot))-\dot{\Lambda}^t(\cdot)\right\|^2 \leqslant \frac{1}{\alpha}\int_0^\alpha \|\dot{\Lambda}^{t+u}(\cdot)-\dot{\Lambda}^t(\cdot)\|^2\,du.$$

But according to (5.3.24)

$$\|\dot{\Lambda}^{t+u}(\cdot)-\dot{\Lambda}^t(\cdot)\|^2 \leqslant u\int_0^u \|\ddot{\Lambda}^{t+\lambda}(\cdot)\|^2\,d\lambda \leqslant \alpha\int_0^\alpha \|\ddot{\Lambda}^{t+\lambda}(\cdot)\|^2\,d\lambda$$

provided $0\leqslant u\leqslant\alpha$ and so we deduce the inequality

$$\left\|\frac{1}{\alpha}(\Lambda^{t+\alpha}(\cdot)-\Lambda^t(\cdot))-\dot{\Lambda}^t(\cdot)\right\|^2 \leqslant \alpha\int_0^\alpha \|\ddot{\Lambda}^{t+u}(\cdot)\|^2\,du$$

from which (5.3.20) follows and the chain-rule is proved.

Before we turn to examining the implications of the Clausius-Duhem inequality for materials with memory we must make an additional mathematical remark. Suppose that the function $\Lambda(\cdot)$ meets the chain-rule conditions, as they were stated above, that t is any time and that $\boldsymbol{\Omega}$ is any element of the vector space $\mathscr{V}$. We assert that it is possible to choose functions $\Lambda_\alpha(\cdot)$, likewise meeting the chain-rule conditions and defined for every sufficiently small positive number α, in such a way that (i) $\Lambda_\alpha(t)=\Lambda(t)$ and $\dot{\Lambda}_\alpha(t)=\boldsymbol{\Omega}$ for every α, and (ii) $\|\Lambda_\alpha^t(\cdot)-\Lambda^t(\cdot)\|\to 0$ and $\|\dot{\Lambda}_\alpha^t(\cdot)-\dot{\Lambda}^t(\cdot)\|\to 0$ as $\alpha\to 0$. In a loose way one can say that the derivative $\dot{\Lambda}(t)$ at t can be chosen arbitrarily without affecting the past histories $\Lambda^t(\cdot)$ and $\dot{\Lambda}^t(\cdot)$, regarded as elements of the Hilbert space $\mathscr{H}$.

One way of constructing the functions $\Lambda_\alpha(s)$ is to introduce a continuous function[1] $p(\cdot)$ having continuous first and second derivatives $\dot{p}(\cdot)$ and $\ddot{p}(\cdot)$ everywhere, vanishing identically for $|s|\geqslant 1$ and with $p(0)=0$, $\dot{p}(0)=1$ and then to define

$$\Lambda_\alpha(s)=\Lambda(s)+\alpha p\left(\frac{s-t}{\alpha}\right)(\boldsymbol{\Omega}-\dot{\Lambda}(t)) \qquad (5.3.26)$$

for every s. Clearly $\Lambda_\alpha(t)=\Lambda(t)$ and $\Lambda_\alpha(t)=\boldsymbol{\Omega}$, that is to say the conditions (i) hold. Furthermore, $\Lambda_\alpha(s)=\Lambda(s)$ for every $s\leqslant t-\alpha$ and for every $s\geqslant t+\alpha$ and it is a routine matter to verify both that $\Lambda_\alpha(\cdot)$ meets all the

[1] The function $p(\cdot)$ defined by $p(s)=0$ for $|s|\geqslant 1$ and $p(s)=s(1-s^2)^3$ for $|s|\leqslant 1$ will do.

hypotheses of the chain-rule, if α is sufficiently small, and that the conditions (ii) hold.

If

$$f_\alpha(t)=\bar{f}(\boldsymbol{\Lambda}_\alpha(t),\boldsymbol{\Lambda}_\alpha^t(\cdot)) \tag{5.3.27}$$

the chain-rule, together with the conditions (i) and (ii) and the continuity assumptions on the functionals $D\bar{f}$ and $\delta\bar{f}$, tells us that

$$\dot{f}_\alpha(t) \to D\bar{f}(\boldsymbol{\Lambda}(t),\boldsymbol{\Lambda}^t(\cdot))\cdot\boldsymbol{\Omega}+\delta\bar{f}(\boldsymbol{\Lambda}(t),\boldsymbol{\Lambda}^t(\cdot)|\dot{\boldsymbol{\Lambda}}^t(\cdot)) \tag{5.3.28}$$

as $\alpha\to 0$. If we write

$$[\dot{f}(t)] = \lim_{\alpha\to 0}\dot{f}_\alpha(t)-\dot{f}(t), \quad [\dot{\boldsymbol{\Lambda}}(t)] = \boldsymbol{\Omega}-\dot{\boldsymbol{\Lambda}}(t) \tag{5.3.29}$$

for the jumps in the derivatives of $f(\cdot)$ and $\boldsymbol{\Lambda}(\cdot)$ at t we have the relation

$$[\dot{f}(t)] = D\bar{f}(\boldsymbol{\Lambda}(t),\boldsymbol{\Lambda}^t(\cdot))\cdot[\dot{\boldsymbol{\Lambda}}(t)], \tag{5.3.30}$$

which can be interpreted to mean that a functional $\bar{f}$ which is continuously differentiable in our sense responds elastically to rapid changes in the rate $\dot{\boldsymbol{\Lambda}}(t)$.

With these preliminaries out of the way we proceed to study the thermodynamics of the materials whose constitutive equations are (5.1.7), (5.1.8), (5.1.9), (5.1.10) and (5.1.11). If we put, as we have done already, $\boldsymbol{\Lambda}(\cdot)=(\boldsymbol{F}(\cdot),\theta(\cdot),\boldsymbol{g}(\cdot))$ these equations can be written as

$$\boldsymbol{T}(t) = \bar{\boldsymbol{T}}(\boldsymbol{\Lambda}(t),\boldsymbol{\Lambda}^t(\cdot)), \tag{5.3.31}$$

$$q(t) = \bar{\boldsymbol{q}}(\boldsymbol{\Lambda}(t),\boldsymbol{\Lambda}^t(\cdot)), \tag{5.3.32}$$

$$e(t) = \bar{e}(\boldsymbol{\Lambda}(t),\boldsymbol{\Lambda}^t(\cdot)), \tag{5.3.33}$$

$$\eta(t) = \bar{\eta}(\boldsymbol{\Lambda}(t),\boldsymbol{\Lambda}^t(\cdot)), \tag{5.3.34}$$

$$\psi(t) = \bar{\psi}(\boldsymbol{\Lambda}(t),\boldsymbol{\Lambda}^t(\cdot)). \tag{5.3.35}$$

We choose to restrict the class of materials being considered by assuming that there is an oblivator such that each of the functionals $\bar{\boldsymbol{T}}$, $\bar{\boldsymbol{q}}$, $\bar{e}$, $\bar{\eta}$ is continuous and the functional $\bar{\psi}$ is continuously differentiable in the sense defined above. Then, provided that $\boldsymbol{\Lambda}(\cdot)$ meets the chain-rule conditions,

$$\begin{aligned}\dot{\psi}(t) &= D\bar{\psi}(\boldsymbol{\Lambda}(t),\boldsymbol{\Lambda}^t(\cdot))\cdot\dot{\boldsymbol{\Lambda}}(t)+\delta\bar{\psi}(\boldsymbol{\Lambda}(t),\boldsymbol{\Lambda}^t(\cdot)|\dot{\boldsymbol{\Lambda}}^t(\cdot))\\ &= \frac{\partial\bar{\psi}}{\partial\boldsymbol{F}}(\boldsymbol{\Lambda}(t),\boldsymbol{\Lambda}^t(\cdot))\cdot\dot{\boldsymbol{F}}(t)\\ &\quad+\frac{\partial\bar{\psi}}{\partial\theta}(\boldsymbol{\Lambda}(t),\boldsymbol{\Lambda}^t(\cdot))\dot{\theta}(t)\\ &\quad+\frac{\partial\bar{\psi}}{\partial\boldsymbol{g}}(\boldsymbol{\Lambda}(t),\boldsymbol{\Lambda}^t(\cdot))\cdot\dot{\boldsymbol{g}}(t)\\ &\quad+\delta\bar{\psi}(\boldsymbol{\Lambda}(t),\boldsymbol{\Lambda}^t(\cdot)|\dot{\boldsymbol{\Lambda}}^t(\cdot))\end{aligned} \tag{5.3.36}$$

where $\partial\bar{\psi}/\partial\boldsymbol{F}$, $\partial\bar{\psi}/\partial\theta$, $\partial\bar{\psi}/\partial\boldsymbol{g}$ are the partial derivatives of the free energy with respect to the present values of the deformation gradient, the temperature and the temperature gradient, the differentiations being performed with the past histories of these quantities being held fixed.

If the chain-rule formula (5.3.36) is substituted into the Clausius-Duhem inequality, in the form (5.1.6), we obtain the inequality

$$\left(\frac{\partial\bar{\psi}}{\partial\boldsymbol{F}}(\boldsymbol{\Lambda}(t),\boldsymbol{\Lambda}^t(\cdot))-\bar{\boldsymbol{S}}(\boldsymbol{\Lambda}(t),\boldsymbol{\Lambda}^t(\cdot))\right)\cdot\dot{\boldsymbol{F}}(t)$$
$$+\left(\frac{\partial\bar{\psi}}{\partial\theta}(\boldsymbol{\Lambda}(t),\boldsymbol{\Lambda}^t(\cdot))+\bar{\eta}(\boldsymbol{\Lambda}(t),\boldsymbol{\Lambda}^t(\cdot))\right)\dot{\theta}(t)+\frac{\partial\bar{\psi}}{\partial\boldsymbol{g}}(\boldsymbol{\Lambda}(t),\boldsymbol{\Lambda}^t(\cdot))\cdot\dot{\boldsymbol{g}}(t)$$
$$+\delta\bar{\psi}(\boldsymbol{\Lambda}(t),\boldsymbol{\Lambda}^t(\cdot)|\dot{\boldsymbol{\Lambda}}^t(\cdot))+\frac{1}{\rho(t)\theta(t)}\boldsymbol{g}(t)\cdot\bar{\boldsymbol{q}}(\boldsymbol{\Lambda}(t),\boldsymbol{\Lambda}^t(\cdot))\leqslant 0, \tag{5.3.37}$$

in which $\bar{\boldsymbol{S}}$ is the response functional for the Piola-Kirchhoff stress. This inequality is to hold in any process $\boldsymbol{\Lambda}(\cdot)$ meeting the chain-rule conditions; in particular it still holds if we replace $\boldsymbol{\Lambda}(\cdot)$ by $\boldsymbol{\Lambda}_\alpha(\cdot)$ as defined in (5.3.26) and with $\boldsymbol{\Omega}=(\boldsymbol{A},a,\boldsymbol{a})$ where $\boldsymbol{A}$ is any tensor, a is any scalar and $\boldsymbol{a}$ is any vector. If we make this replacement and then take the limit as $\alpha\to 0$ we are left with the inequality

$$\left(\frac{\partial\bar{\psi}}{\partial\boldsymbol{F}}(\boldsymbol{\Lambda}(t),\boldsymbol{\Lambda}^t(\cdot))-\bar{\boldsymbol{S}}(\boldsymbol{\Lambda}(t),\boldsymbol{\Lambda}^t(\cdot))\right)\cdot\boldsymbol{A}$$
$$+\left(\frac{\partial\bar{\psi}}{\partial\theta}(\boldsymbol{\Lambda}(t),\boldsymbol{\Lambda}^t(\cdot))+\bar{\eta}(\boldsymbol{\Lambda}(t),\boldsymbol{\Lambda}^t(\cdot))\right)a+\frac{\partial\bar{\psi}}{\partial\boldsymbol{g}}(\boldsymbol{\Lambda}(t),\boldsymbol{\Lambda}^t(\cdot))\cdot\boldsymbol{a}$$
$$+\delta\bar{\psi}(\boldsymbol{\Lambda}(t),\boldsymbol{\Lambda}^t(\cdot)|\dot{\boldsymbol{\Lambda}}^t(\cdot))+\frac{1}{\rho(t)\theta(t)}\boldsymbol{g}(t)\cdot\bar{\boldsymbol{q}}(\boldsymbol{\Lambda}(t),\boldsymbol{\Lambda}^t(\cdot))\leqslant 0,$$

which can hold for all choices of the tensor $\boldsymbol{A}$, the scalar a and the vector $\boldsymbol{a}$ if and only if

I the Piola-Kirchhoff stress is determined by the free energy through the relation

$$\bar{\boldsymbol{S}}=\frac{\partial\bar{\psi}}{\partial\boldsymbol{F}}, \tag{5.3.38}$$

II the entropy is determined by the free energy through the relation

$$\bar{\eta}=-\frac{\partial\bar{\psi}}{\partial\theta}, \tag{5.3.39}$$

III the free energy is independent of the present value of the temperature gradient in the sense that

$$\frac{\partial \bar{\psi}}{\partial \boldsymbol{g}} = \mathbf{0}, \tag{5.3.40}$$

and

IV the generalised dissipation inequality

$$\delta\bar{\psi} + \frac{1}{\rho\theta}\boldsymbol{g}\cdot\bar{\boldsymbol{q}} \leqslant 0 \tag{5.3.41}$$

holds.

It is easily checked that *the conditions I, II, III and IV are not merely necessary for the Clausius-Duhem inequality* (5.3.37) *to hold, they are also sufficient.* They were all derived by Coleman in his original paper [8].

Clearly the materials considered here have the property that the entire stress response and the entire entropy response, and not just the equilibrium stress response and the equilibrium response, are determined by the free energy. Indeed (5.3.38) tells us that the symmetric stress tensor is given by

$$\bar{\boldsymbol{T}} = \rho\frac{\partial \bar{\psi}}{\partial \boldsymbol{F}}\boldsymbol{F}^T = \frac{\rho_0}{\det \boldsymbol{F}}\frac{\partial \bar{\psi}}{\partial \boldsymbol{F}}\boldsymbol{F}^T. \tag{5.3.42}$$

Because of (5.3.40) the stress and the entropy are independent of the present values of the temperature gradient. In his original paper Coleman assumed from the outset that the constitutive relations involve the temperature gradient only through its present value so that equations (5.3.1), (5.3.2), (5.3.3), (5.3.4) and (5.3.5) are the appropriate ones. In that case the free energy, the stress and the entropy turn out to be independent of the temperature gradient altogether and if we set $\boldsymbol{g}=\mathbf{0}$ in the generalised dissipation inequality (5.3.41) we obtain *the internal dissipation inequality*

$$\delta\bar{\psi} \leqslant 0. \tag{5.3.43}$$

It does not necessarily follow from (5.3.41) and (5.3.43) that the heat conduction inequality

$$\boldsymbol{g}\cdot\boldsymbol{q} \leqslant 0$$

holds but, as we shall see in section 5.4, this inequality does hold in equilibrium, when the histories of the deformation gradient and the temperature are constant.

It must be remembered that although the class of materials meeting the fading memory hypotheses of this section is extensive it is by no means an all-embracing one. As we have seen, a fading memory hypothesis which is formulated in terms of an oblivator implies that in a certain sense the material responds elastically to rapid changes of the deformation gradient, the temperature and the temperature gradient. Materials of the differential type, other than thermoelastic materials, certainly do not satisfy the hypotheses even though they do have fading memory in the sense of section 3.2. For them the result *I* does not hold; it is only the equilibrium stress which is determined by the free energy.

It is perhaps an open question whether the Clausius-Duhem inequality is an appropriate expression of the second law or not but leaving that question aside it is clear that by comparison with most earlier writings on thermodynamics[1] Coleman's theory is a remarkable advance in that it sets out from a clearly defined starting point, namely the Clausius-Duhem inequality, and it uses only concepts which are precise enough to withstand translation into mathematical terms and arrives at definite results of considerable generality.

5.4 Behaviour near Equilibrium

A number of important consequences shedding light on the behavious of materials with memory at or near to equilibrium can be derived from the results *I*, *II*, *III* and *IV* of the preceding section. For the sake of simplicity[2] we shall consider only materials of the kind considered by Coleman in his original paper [8]; this means, as we have seen, that the free energy, the stress, the entropy and the internal energy are independent of the temperature gradient and that the heat flux vector depends on the temperature gradient only through its present value. For this reason we now find it convenient to make a slight change in notation and write $\Lambda(\cdot)=(\boldsymbol{F}(\cdot),\theta(\cdot))$ and to let $\mathscr{V}$ be the vector space of all pairs $(\boldsymbol{A},a)$ consisting of a tensor $\boldsymbol{A}$ and a scalar a and to let $\mathscr{U}$ be the open and connected subset of pairs with $\det\boldsymbol{A}>0$ and $a>0$.

If $\boldsymbol{\Omega}=(\boldsymbol{F}_0,\theta_0)$ is any fixed element of $\mathscr{U}$ we shall write $\boldsymbol{T}^*(\boldsymbol{\Omega})$, $e^*(\boldsymbol{\Omega})$, $\eta^*(\boldsymbol{\Omega})$, $\psi^*(\boldsymbol{\Omega})$ for the stress, the internal energy, the entropy and the free energy arising from the constant process with $\Lambda(s)=\boldsymbol{\Omega}$, that is with $\boldsymbol{F}(s)=\boldsymbol{F}_0$, $\theta(s)=\theta_0$ for every s. In other words if we write the constitutive equations (5.3.1), (5.3.2), (5.3.3), (5.3.4) and (5.3.5) in the form

---

[1] Notable exceptions to the low standards of clarity and precision prevailing in writing on classical thermodynamics are provided by Truesdell's account in the Mechanical Foundations [66] and by Truesdell and Toupin's account in the Classical Field Theories [69].

[2] For a more general definition of equilibrium, applicable to materials in which dependence upon the past history of the temperature gradient is allowed, the reader is referred to Gurtin [48] and to the account of Gurtin's work in Truesdell [67].

$$T(t) = \bar{T}(\Lambda(t), \Lambda^t(\cdot)), \tag{5.4.1}$$

$$q(t) = \bar{q}(\Lambda(t), \Lambda^t(\cdot), g(t)), \tag{5.4.2}$$

$$e(t) = \bar{e}(\Lambda(t), \Lambda^t(\cdot)), \tag{5.4.3}$$

$$\eta(t) = \bar{\eta}(\Lambda(t), \Lambda^t(\cdot)), \tag{5.4.4}$$

$$\psi(t) = \bar{\psi}(\Lambda(t), \Lambda^t(\cdot)) \tag{5.4.5}$$

and if Ω^* is the constant past history whose value is Ω at all previous times then

$$\bar{T}^*(\Omega) = \bar{T}(\Omega, \Omega^*), \quad e^*(\Omega) = \bar{e}(\Omega, \Omega^*), \quad \eta^*(\Omega) = \bar{\eta}(\Omega, \Omega^*), \quad \psi^*(\Omega) = \bar{\psi}(\Omega, \Omega^*) \tag{5.4.6}$$

and $T^*(\cdot)$, $e^*(\cdot)$, $\eta^*(\cdot)$, $\psi^*(\cdot)$ are just functions defined on $\mathscr{U}$; they are the response functions for the stress, the internal energy, the entropy and the free energy in equilibrium.

We begin by showing that the free energy has a certain relaxation property under constant continuation of a given process. If $\Lambda(\cdot) = (F(\cdot), \theta(\cdot))$ is a process and t is any fixed time then by the *constant continuation* at time t is meant the process $\hat{\Lambda}(s)$ coinciding with $\Lambda(s)$ at all times prior to t and then held constant subsequently, that is $\hat{\Lambda}(s) = \Lambda(s)$ for $s \leqslant t$ and $\hat{\Lambda}(s) = \Lambda(t)$ for $s \geqslant t$. We shall show that under certain conditions the past history $\hat{\Lambda}^u(\cdot)$ lies in the Hilbert space $\mathscr{H}$ for every $u > t$ and that if $\Lambda(t)^*$ is the constant past history whose value is always $\Lambda(t)$ then

$$\|\hat{\Lambda}^u(\cdot) - \Lambda(t)^*\| \to 0 \tag{5.4.7}$$

as $u \to +\infty$. These facts, together with the continuity of the free energy, then imply that as $u \to +\infty$

$$\psi(u) = \bar{\psi}(\hat{\Lambda}(u), \hat{\Lambda}^u(\cdot)) \to \bar{\psi}(\Lambda(t), \Lambda(t)^*) = \psi^*(\Lambda(t)), \tag{5.4.8}$$

that is to say *under constant continuation the free energy relaxes to the equilibrium value* $\psi^*(\Lambda(t))$, as we should expect in a material with fading memory. Exactly similar remarks apply to the stress, the internal energy and the entropy.

The fact that $\hat{\Lambda}^u(\cdot)$ is in $\mathscr{H}$ can be proved very quickly: for

$$\begin{aligned}
\|\hat{\Lambda}^u(\cdot)\|^2 &= \int_0^{+\infty} \gamma(s)^2 |\hat{\Lambda}(u-s)|^2 \, ds \\
&= \int_0^{u-t} \gamma(s)^2 |\Lambda(t)|^2 \, ds + \int_{u-t}^{+\infty} \gamma(s)^2 |\hat{\Lambda}(u-s)|^2 \, ds \\
&= \int_0^{u-t} \gamma(s)^2 |\Lambda(t)|^2 \, ds + \int_0^{+\infty} \gamma(s-t+u)^2 |\Lambda^t(s)|^2 \, ds.
\end{aligned}$$

Because the oblivator $\gamma(\cdot)$ is monotone decreasing $\gamma(s-t+u)^2 \leqslant \gamma(s)^2$ if $u>t$ and so

$$\|\hat{\Lambda}^u(\cdot)\|^2 \leqslant |\Lambda(t)|^2 \int_0^{+\infty} \gamma(s)^2\, ds + \|\Lambda^t(\cdot)\|^2,$$

which proves that $\hat{\Lambda}^u(\cdot)$ is in $\mathscr{H}$ if $\Lambda^t(\cdot)$ is. Again, if we use the inequality $(\alpha-\beta)^2 \leqslant 2(\alpha^2+\beta^2)$ we find that

$$\begin{aligned}\|\hat{\Lambda}^u(\cdot)-\Lambda(t)^*\|^2 &= \int_0^{+\infty} \gamma(s)^2 |\hat{\Lambda}(u-s)-\Lambda(t)|^2\, ds \\ &= \int_{u-t}^{+\infty} \gamma(s)^2 |\Lambda(u-s)-\Lambda(t)|^2\, ds \\ &\leqslant 2 \int_{u-t}^{+\infty} \gamma(s)^2 |\Lambda(u-s)|^2\, ds + 2|\Lambda(t)|^2 \int_{u-t}^{+\infty} \gamma(s)^2\, ds,\end{aligned}$$

that is

$$\|\hat{\Lambda}^u(\cdot)-\Lambda(t)^*\|^2 \leqslant 2\int_0^{+\infty} \gamma(s-t+u)^2 |\Lambda^t(s)|^2\, ds + 2|\Lambda(t)|^2 \int_{u-t}^{+\infty} \gamma(s)^2\, ds. \tag{5.4.9}$$

Since $\gamma(\cdot)$ decreases monotonely

$$\gamma(s-t+u)^2 |\Lambda^t(s)|^2 \leqslant \gamma(s)^2 |\Lambda^t(s)|^2$$

if $u>t$. The right-hand side of this inequality is an integrable function of s and, for each fixed s, $\gamma(s-t+u)\to 0$ as $u\to+\infty$ and so the dominated convergence theorem of Lebesgue implies that

$$\int_0^{+\infty} \gamma(s-t+u)^2 |\Lambda^t(s)|^2\, ds \to 0$$

as $u\to+\infty$. Furthermore

$$\int_{u-t}^{+\infty} \gamma(s)^2\, ds \to 0$$

as $u\to+\infty$ because $\gamma(\cdot)$ is square integrable and the required result (5.4.7) follows from the inequality (5.4.9).

Because of the relations (5.3.38) and (5.3.39) proved in the preceding section the chain-rule (5.3.15) for the free energy can be written in the form

$$\dot{\psi} = \boldsymbol{S}\cdot\dot{\boldsymbol{F}} - \eta\dot{\theta} + \delta\overline{\psi} \tag{5.4.10}$$

and because of the internal dissipation inequality (5.3.43) the inequality

$$\dot{\psi} \leqslant \boldsymbol{S}\cdot\dot{\boldsymbol{F}} - \eta\dot{\theta} \tag{5.4.11}$$

holds in any process $\Lambda(\cdot)$ meeting the chain-rule conditions. In particular if $\dot{\boldsymbol{F}}=\boldsymbol{0}$ and $\dot{\theta}=0$ then $\dot{\psi}\leqslant 0$; in other words *if the deformation gradient and the temperature are held fixed the free energy cannot increase.*

Even if the process $\boldsymbol{\Lambda}(\cdot)$ meets the chain-rule conditions its constant continuation $\hat{\boldsymbol{\Lambda}}(\cdot)$ at t cannot be expected to meet them, for it usually fails to be twice continuously differentiable at t, and the result just proved is not immediately applicable to $\hat{\boldsymbol{\Lambda}}(\cdot)$. However the discontinuity at t can be smoothed out by introducing any twice continuously differentiable bridging function $q(\cdot)$ which is monotone increasing and which is so chosen that $q(s)\equiv 0$ for $s\leqslant 0$ and $q(s)\equiv 1$ for $s\geqslant 1$ and setting

$$\boldsymbol{\Lambda}_\varepsilon(s)=\left(1-q\left(\frac{s-t}{\varepsilon}\right)\right)\boldsymbol{\Lambda}(s)+q\left(\frac{s-t+\varepsilon}{\varepsilon}\right)\boldsymbol{\Lambda}(t)$$

for each small positive ε. Then $\boldsymbol{\Lambda}_\varepsilon(s)\equiv\boldsymbol{\Lambda}(s)$ for $s\leqslant t-\varepsilon$, $\boldsymbol{\Lambda}_\varepsilon(s)=\boldsymbol{\Lambda}(t)$ for $s\geqslant t$ and $\boldsymbol{\Lambda}_\varepsilon(\cdot)$ does meet the chain-rule conditions and so, as we have seen,

$$\frac{d}{du}\bar{\psi}(\boldsymbol{\Lambda}_\varepsilon(u),\boldsymbol{\Lambda}_\varepsilon^u(\cdot))\leqslant 0$$

for every $u\geqslant t$. This fact, together with the relaxation property (5.4.8), yields as a consequence the inequality

$$\bar{\psi}(\boldsymbol{\Lambda}_\varepsilon(t),\boldsymbol{\Lambda}_\varepsilon^t(\cdot))\geqslant\psi^*(\boldsymbol{\Lambda}_\varepsilon(t)). \tag{5.4.12}$$

But $\boldsymbol{\Lambda}_\varepsilon(t)=\boldsymbol{\Lambda}(t)$ and, as can easily be verified, the past histories $\boldsymbol{\Lambda}_\varepsilon^t(\cdot)$ approximate $\boldsymbol{\Lambda}^t(\cdot)$ in the sense that $\|\boldsymbol{\Lambda}_\varepsilon^t(\cdot)-\boldsymbol{\Lambda}^t(\cdot)\|\to 0$ as $\varepsilon\to 0$ and thus if we take the limit as $\varepsilon\to 0$ in (5.4.12) we are left with the inequality

$$\bar{\psi}(\boldsymbol{\Lambda}(t),\boldsymbol{\Lambda}^t(\cdot))\geqslant\psi^*(\boldsymbol{\Lambda}(t)) \tag{5.4.13}$$

asserting that *among all histories ending with given values of the deformation gradient and the temperature the constant history yields the least free energy.*

The derivation of the minimal property of the free energy expressed by the inequality (5.4.13) is one of the major achievements of Coleman's theory. In a series of papers, Coleman and Greenberg [12], Coleman and Dill [11], Coleman and Mizel [23, 24] and Coleman [10] have shown that this property has important implications for stability theory. More specifically, they have used it to provide a rigorous justification for an assumption commonly made in physics, namely that any equilibrium state of a thermodynamic system at which the equilibrium free energy $\psi^*(\cdot)$ has a strict local minimum is in fact dynamically stable in an appropriate sense.

On integrating both sides of the inequality (5.4.11) between any two times t_0 and t_1 $(t_1>t_0)$ we obtain the inequality

$$\int_{t_0}^{t_1}(\boldsymbol{S}(t)\cdot\dot{\boldsymbol{F}}(t)-\eta(t)\dot{\theta}(t))\,dt\geqslant\psi(t_1)-\psi(t_0) \tag{5.4.14}$$

where, according to (5.4.13),

$$\psi(t_1) \geqslant \psi^*(\boldsymbol{F}(t_1), \theta(t_1)).$$

If the deformation gradient and the temperature happen to be constant at all times before t_0 the free energy at that time is

$$\psi(t_0) = \psi^*(\boldsymbol{F}(t_0), \theta(t_0))$$

and so the inequality (5.4.14) tells us in this case that

$$\int_{t_0}^{t_1} (\boldsymbol{S}(t) \cdot \dot{\boldsymbol{F}}(t) - \eta(t)\dot{\theta}(t)\, dt \geqslant \psi^*(\boldsymbol{F}(t_1), \theta(t_1)) - \psi^*(\boldsymbol{F}(t_0), \theta(t_0)). \tag{5.4.15}$$

In particular if the process is isothermal and if the deformation gradient returns to its original value at the time t_1, that is if $\dot{\theta}(\cdot) = 0$ and $\boldsymbol{F}(t_0) = \boldsymbol{F}(t_1)$, then

$$\int_{t_0}^{t_1} \boldsymbol{S}(t) \cdot \dot{\boldsymbol{F}}(t)\, dt \geqslant 0 \tag{5.4.16}$$

which means that *the work done around any closed isothermal path starting from equilibrium cannot be negative.*

The minimal property of the free energy expressed by the inequality (5.4.13) can be made the basis for deriving the counterparts to the relations (5.3.38) and (5.3.39) in thermostatic equilibrium. To derive them one can proceed in the following way. Suppose that $\boldsymbol{\Omega}$ is any fixed element of $\mathscr{U}$ and that $\boldsymbol{\Gamma}$ is any element of $\mathscr{V}$ such that the line segment $\boldsymbol{\Omega} + s\boldsymbol{\Gamma}$, $0 \leqslant s \leqslant 1$, lies in $\mathscr{U}$. If we introduce the process

$$\boldsymbol{\Lambda}_\varepsilon(s) = \boldsymbol{\Omega} + q\left(\frac{s+\varepsilon}{\varepsilon}\right)\boldsymbol{\Gamma}, \tag{5.4.17}$$

where ε is any small positive number and q is the bridging function introduced earlier, then $\boldsymbol{\Lambda}_\varepsilon(\cdot)$ meets the chain-rule conditions and $\boldsymbol{\Lambda}_\varepsilon(s) = \boldsymbol{\Omega}$ for every $s \leqslant -\varepsilon$ and $\boldsymbol{\Lambda}_\varepsilon(s) = \boldsymbol{\Omega} + \boldsymbol{\Gamma}$ for every $s \geqslant 0$. According to (5.4.13) the free energy at $s = 0$ satisfies the inequality

$$\bar{\psi}(\boldsymbol{\Omega} + \boldsymbol{\Gamma}, \boldsymbol{\Lambda}_\varepsilon^0(\cdot)) \geqslant \psi^*(\boldsymbol{\Omega} + \boldsymbol{\Gamma}).$$

However, it is not difficult to show that as $\varepsilon \to 0$ the past histories $\boldsymbol{\Lambda}_\varepsilon^0(\ )$ approximate the constant history $\boldsymbol{\Omega}^*$ in the usual way, namely $\|\boldsymbol{\Lambda}_\varepsilon^0(\ ) - \boldsymbol{\Omega}^*\| \to 0$ and because of this fact we have

$$\bar{\psi}(\boldsymbol{\Omega} + \boldsymbol{\Gamma}, \boldsymbol{\Omega}^*) \geqslant \psi^*(\boldsymbol{\Omega} + \boldsymbol{\Gamma}) \tag{5.4.18}$$

for every $\boldsymbol{\Gamma}$ sufficiently close to $\boldsymbol{0}$. For each fixed $\boldsymbol{\Omega}$ then,

$$\bar{\psi}(\boldsymbol{\Omega} + \boldsymbol{\Gamma}, \boldsymbol{\Omega}^*) - \psi^*(\boldsymbol{\Omega} + \boldsymbol{\Gamma})$$

is a non-negative function of $\boldsymbol{\Gamma}$ vanishing when $\boldsymbol{\Gamma}=\mathbf{0}$ and consequently its gradient taken with respect to $\boldsymbol{\Gamma}$ vanishes at $\boldsymbol{\Gamma}=\mathbf{0}$, that is

$$\frac{\partial\bar{\psi}}{\partial\boldsymbol{\Lambda}}(\boldsymbol{\Omega},\boldsymbol{\Omega}^*)=\frac{\partial\psi^*}{\partial\boldsymbol{\Lambda}}(\boldsymbol{\Omega}). \tag{5.4.19}$$

Bearing in mind that $\boldsymbol{\Lambda}$ in $\mathscr{U}$ is a pair $(\boldsymbol{F},\theta)$, equation (5.4.19) states that

$$\frac{\partial\bar{\psi}}{\partial\boldsymbol{F}}(\boldsymbol{\Omega},\boldsymbol{\Omega}^*)=\frac{\partial\psi^*}{\partial\boldsymbol{F}}(\boldsymbol{\Omega}),\qquad \frac{\partial\bar{\psi}}{\partial\theta}(\boldsymbol{\Omega},\boldsymbol{\Omega}^*)=\frac{\partial\psi^*}{\partial\theta}(\boldsymbol{\Omega})$$

and we deduce immediately from these relations and from (5.3.38) and (5.3.39) the thermostatic relations

$$\boldsymbol{S}^*=\frac{\partial\psi^*}{\partial\boldsymbol{F}}, \tag{5.4.20}$$

$$\eta^*=-\frac{\partial\psi^*}{\partial\theta} \tag{5.4.21}$$

which show that *the equilibrium free energy determines the equilibrium stress and the equilibrium entropy through formulae agreeing with those found previously in section* 3.3.

Incidentally the argument leading to (5.4.19) has shown that with the stated conditions on $\boldsymbol{\Omega}$ and $\boldsymbol{\Gamma}$

$$\delta\bar{\psi}(\boldsymbol{\Omega},\boldsymbol{\Omega}^*\,|\,\boldsymbol{\Gamma}^*)=0. \tag{5.4.22}$$

Now that we have arrived at the classical formulae (5.4.20) and (5.4.21) appropriate to thermostatic equilibrium we can use the idea of a retardation of a process, an idea which has already been used in section 3.3, to display (5.4.20) and (5.4.21) as the asymptotic limits of the relations (5.3.38) and (5.3.39) obtained when a given process is retarded sufficiently. We confine our attention to processes $\boldsymbol{\Lambda}(\cdot)=(\boldsymbol{F}(\cdot),\theta(\cdot))$ meeting the chain-rule conditions and for the sake of simplicity we suppose that $\|\boldsymbol{\Lambda}(s)\|$ and $\|\dot{\boldsymbol{\Lambda}}(s)\|$ remain bounded, at least for times $s\leqslant t$, where t is to be fixed throughout the discussion. The *retardation* $\boldsymbol{\Lambda}_\alpha(\cdot)$ of $\boldsymbol{\Lambda}(\cdot)$ at time t is the process obtained by replacing the standard time-scale s by the retarded time-scale $t+\alpha(s-t)$, where α lying in $0<\alpha<1$ is the retardation. Thus

$$\boldsymbol{\Lambda}_\alpha(s)=\boldsymbol{\Lambda}(t+\alpha(s-t)) \tag{5.4.23}$$

and $\boldsymbol{\Lambda}_\alpha(t)=\boldsymbol{\Lambda}(t)$. We might expect that as the process is retarded more and more the past history of the retarded process up to the time t would approximate to the constant history with the value $\boldsymbol{\Lambda}(t)$. This is so; indeed

$$\|\boldsymbol{\Lambda}_\alpha^t(\cdot)-\boldsymbol{\Lambda}(t)^*\|\to 0 \tag{5.4.24}$$

as $\alpha \to 0$. To prove that (5.4.24) does hold we need only observe that

$$\|\Lambda_\alpha^t(\cdot) - \Lambda(t)^*\|^2 = \int_0^{+\infty} \gamma(s)^2 |\Lambda_\alpha(t-s) - \Lambda(t)|^2 \, ds$$

$$= \int_0^{+\infty} \gamma(s)^2 |\Lambda(t-\alpha s) - \Lambda(t)|^2 \, ds,$$

that, because we have assumed that $|\Lambda(s)|$ is bounded on $s \leqslant t$,

$$\gamma(s)^2 |\Lambda(t-\alpha s) - \Lambda(t)|^2 \leqslant \text{constant} \times \gamma(s)^2$$

and that, because of the continuity of $\Lambda(s)$,

$$\gamma(s)^2 |\Lambda(t-\alpha s) - \Lambda(t)|^2 \to 0$$

as $\alpha \to 0$ and then apply the dominated convergence theorem of Lebesgue. The derivative of the retarded process is

$$\dot{\Lambda}_\alpha(s) = \alpha \dot{\Lambda}(t + \alpha(s-t)),$$

its past history up to t is

$$\dot{\Lambda}_\alpha^t(s) = \alpha \dot{\Lambda}(t - \alpha s) = \alpha \dot{\Lambda}^t(\alpha s)$$

and it can be shown, in a similar way to (5.4.24), that

$$\frac{1}{\alpha} \|\dot{\Lambda}_\alpha^t(\cdot) - \alpha \dot{\Lambda}(t)^*\| \to 0 \tag{5.4.25}$$

as $\alpha \to 0$.

Because of (5.4.24) it follows, as we should expect, that the stress, the internal energy, the entropy and the gree energy approach the equilibrium values

$$T_\alpha(t) = \bar{T}(\Lambda_\alpha(t), \Lambda_\alpha^t(\cdot)) \to T^*(\Lambda(t)), \tag{5.4.26}$$

$$e_\alpha(t) = \bar{e}(\Lambda_\alpha(t), \Lambda_\alpha^t(\cdot)) \to e^*(\Lambda(t)), \tag{5.4.27}$$

$$\eta_\alpha(t) = \bar{\eta}(\Lambda_\alpha(t), \Lambda_\alpha^t(\cdot)) \to \eta^*(\Lambda(t)), \tag{5.4.28}$$

$$\psi_\alpha(t) = \bar{\psi}(\Lambda_\alpha(t), \Lambda_\alpha^t(\cdot)) \to \psi^*(\Lambda(t)) \tag{5.4.29}$$

as the retardation $\alpha \to 0$.

For the retarded process the form (5.4.10) of the chain-rule for the free energy becomes the statement

$$\dot{\psi}_\alpha(t) = \bar{S}(\Lambda_\alpha(t), \Lambda_\alpha^t(\cdot)) \cdot \dot{F}_\alpha(t) - \bar{\eta}(\Lambda_\alpha(t), \Lambda_\alpha^t(\cdot)) \dot{\theta}_\alpha(t) + \delta\bar{\psi}(\Lambda_\alpha(t), \Lambda_\alpha^t(\cdot) \,|\, \dot{\Lambda}_\alpha^t(\cdot)) \tag{5.4.30}$$

The properties of the functional $\delta\bar{\psi}$ and equations (5.4.24) and (5.4.25) when taken together imply that

$$\frac{1}{\alpha} \{\delta\bar{\psi}(\Lambda_\alpha(t), \Lambda_\alpha^t(\cdot) \,|\, \dot{\Lambda}_\alpha^t(\cdot)) - \delta\bar{\psi}(\Lambda(t), \Lambda(t)^* \,|\, \dot{\Lambda}(t)^*)\} \to 0$$

as $\alpha \to 0$. However it has been shown already, and the result set down as equation (5.4.22), that

$$\delta\bar{\psi}(\boldsymbol{\Lambda}(t), \boldsymbol{\Lambda}(t)^* \,|\, \dot{\boldsymbol{\Lambda}}(t)^*) = 0$$

and thus

$$\frac{1}{\alpha}\,\delta\bar{\psi}(\boldsymbol{\Lambda}_\alpha(t), \boldsymbol{\Lambda}^t_\alpha(\cdot) \,|\, \dot{\boldsymbol{\Lambda}}^t_\alpha(\cdot)) \to 0 \tag{5.4.31}$$

as $\alpha \to 0$. Since the entropy satisfies (5.4.28) and the Piola-Kirchhoff stress satisfies

$$\bar{\boldsymbol{S}}(\boldsymbol{\Lambda}_\alpha(t), \boldsymbol{\Lambda}^t_\alpha(\cdot)) \to \boldsymbol{S}^*(\boldsymbol{\Lambda}(t))$$

as $\alpha \to 0$ and since $\dot{\boldsymbol{F}}_\alpha(t) = \alpha \dot{\boldsymbol{F}}(t)$, $\dot{\theta}_\alpha(t) = \alpha \dot{\theta}(t)$ it follows from (5.4.30) and (5.4.31) that

$$\frac{1}{\alpha}\{\dot{\psi}_\alpha(t) - \boldsymbol{S}^*(t) \cdot \dot{\boldsymbol{F}}_\alpha(t) + \eta^*(t)\dot{\theta}_\alpha(t)\} \to 0 \tag{5.4.32}$$

as $\alpha \to 0$, where we have written $\boldsymbol{S}^*(t)$ for $\boldsymbol{S}^*(\boldsymbol{\Lambda}(t))$ and $\eta^*(t)$ for $\eta^*(\boldsymbol{\Lambda}(t))$. The fact just proved amounts to saying that *in the retardation of a fixed process by an amount* α *the Gibbs relation*

$$\dot{\psi}_\alpha(t) = \boldsymbol{S}^*(t) \cdot \dot{\boldsymbol{F}}_\alpha(t) - \eta^*(t)\dot{\theta}_\alpha(t) \tag{5.4.33}$$

holds to within an error whose order of magnitude is $o(\alpha)$. As Coleman has remarked[1], it is precisely because the Gibbs relation holds asymptotically in this sense that it can be used in linear irreversible thermodynamics and in the classical hydrodynamics of compressible viscous fluids.

Lastly let us look at heat conduction near equilibrium. If $(\boldsymbol{F}_0, \theta_0)$ is any element of $\mathscr{U}$ then

$$\delta\bar{\psi}(\boldsymbol{F}_0, \theta_0, \boldsymbol{F}_0^*, \theta_0^* \,|\, \boldsymbol{0}^*) = 0$$

and so if we consider the constant process with $(\boldsymbol{F}(s), \theta(s), \boldsymbol{g}(s)) = (\boldsymbol{F}_0, \theta_0, \boldsymbol{g})$ for every s, where $\boldsymbol{g}$ is any vector, it follows from the generalised dissipation inequality (5.3.41), or directly from the Clausius-Duhem inequality (5.1.6), that the *heat conduction inequality*

$$\boldsymbol{g} \cdot \bar{\boldsymbol{q}}(\boldsymbol{F}_0, \theta_0, \boldsymbol{F}_0^*, \theta_0^*, \boldsymbol{g}) \leqslant 0 \tag{5.4.34}$$

holds. Once again we can deduce that there is no piezo-caloric effect in equilibrium:

$$\bar{\boldsymbol{q}}(\boldsymbol{F}_0, \theta_0, \boldsymbol{F}_0^*, \theta_0^*, \boldsymbol{0}) = \boldsymbol{0}, \tag{5.4.35}$$

---

[1] See § 10 of [8].

that the conductivity tensor in equilibrium

$$K(F_0,\theta_0) = -\frac{\partial \bar{q}}{\partial g}(F_0,\theta_0,F_0^*,\theta_0^*,g)\bigg|_{g=0} \tag{5.4.36}$$

is positive semi-definite and that Fourier's law holds near equilibrium in the sense that

$$\bar{q}(F_0,\theta_0,F_0^*,\theta_0^*,g) = -K(F_0,\theta_0)g + o(g). \tag{5.4.37}$$

5.5 The Internal Energy as an Independent Variable

The results which were proved in sections 5.3 and 5.4 can all be rephrased if we use the deformation gradient and the internal energy as independent variables instead of the deformation gradient and the temperature; many of them then take on forms familiar to us from the discussion given in Chapters 2, 3 and 4.

When it is written out in full and the terms rearranged the definition (5.1.4) of the free energy states that

$$e(t) = \bar{\psi}(F(t),\theta(t),F^t(\cdot),\theta^t(\cdot)) + \theta(t)\bar{\eta}(F(t),\theta(t),F^t(\cdot),\theta^t(\cdot)). \tag{5.5.1}$$

If (5.5.1) can be inverted uniquely to express the temperature as a functional

$$\theta(t) = \breve{\theta}(F(t),e(t),F^t(\cdot),e^t(\cdot)) \tag{5.5.2}$$

of the deformation gradient and internal energy histories and if we substitute (5.5.2) into (5.4.1), (5.4.2) and (5.4.4) we are left with constitutive relations of the form

$$T(t) = \breve{T}(F(t),e(t),F^t(\cdot),e^t(\cdot)), \tag{5.5.3}$$

$$q(t) = \breve{q}(F(t),e(t),F^t(\cdot),e^t(\cdot),g(t)), \tag{5.5.4}$$

$$\eta(t) = \breve{\eta}(F(t),e(t),F^t(\cdot),e^t(\cdot)) \tag{5.5.5}$$

for the stress, the heat flux and the entropy.

Once we have the constitutive relations in the forms (5.5.2), (5.5.3), (5.5.4) and (5.5.5) it is possible to proceed on lines substantially parallel to those of sections 5.3 and 5.4. To start with we take it that the past history $(F^t(\cdot),e^t(\cdot))$ lies always in a Hilbert space, defined in terms of an oblivator as before, and that the response functionals $\breve{\theta}$, $\breve{T}$, $\breve{q}$ are continuous, in the fading memory sense, whilst $\breve{\eta}$ is continuously differentiable so that a chain-rule applies in the calculation of the derivative $\dot{\eta}$:

$$\dot{\eta} = \frac{\partial \breve{\eta}}{\partial F}\cdot\dot{F} + \frac{\partial \breve{\eta}}{\partial \theta}\dot{\theta} + \delta\breve{\eta}. \tag{5.5.6}$$

The Clausius-Duhem inequality tells us that

$$\dot{\eta} \geqslant \frac{1}{\theta}\dot{e} - \frac{1}{\theta}\boldsymbol{S}\cdot\dot{\boldsymbol{F}} + \frac{1}{\rho\theta^2}\boldsymbol{g}\cdot\boldsymbol{q} \tag{5.5.7}$$

and when we are using the internal energy as an independent variable it is convenient to work with this inequality rather than with the equivalent inequality (5.1.6) used in section 5.3. Arguments of the kind used before lead us to conclude that *the inequality* (5.5.7) *can hold in every process meeting the chain-rule conditions if and only if*

I the stress is determined by the entropy through the relation

$$\breve{S} = \frac{-\dfrac{\partial\breve{\eta}}{\partial \boldsymbol{F}}}{\dfrac{\partial\breve{\eta}}{\partial e}}, \tag{5.5.8}$$

II the temperature is determined by the entropy through the relation

$$\breve{\theta} = \frac{1}{\dfrac{\partial\breve{\eta}}{\partial e}}, \tag{5.5.9}$$

III the generalised dissipation inequality

$$\delta\breve{\eta} \geqslant \frac{1}{\rho\theta^2}\boldsymbol{g}\cdot\boldsymbol{q} \tag{5.5.10}$$

holds.

Because of (5.5.8) the symmetric stress tensor is

$$\breve{\boldsymbol{T}} = \frac{-\rho\dfrac{\partial\breve{\eta}}{\partial\boldsymbol{F}}\boldsymbol{F}^T}{\dfrac{\partial\breve{\eta}}{\partial e}} = \frac{-\rho_0\dfrac{\partial\breve{\eta}}{\partial\boldsymbol{F}}\boldsymbol{F}^T}{\det\boldsymbol{F}\dfrac{\partial\breve{\eta}}{\partial e}} \tag{5.5.11}$$

If we set $\boldsymbol{g}=\boldsymbol{0}$ in (5.5.10) we obtain the *internal dissipation inequality*

$$\delta\breve{\eta} \geqslant 0 \tag{5.5.12}$$

which is the counterpart of (5.3.43).

Corresponding to the minimal property of the free energy expressed by the inequality (5.4.13) the entropy has a maximal property, namely

$$\breve{\eta}(\boldsymbol{F}(t), e(t), \boldsymbol{F}^t(\cdot), e^t(\cdot)) \leqslant \eta^*(\boldsymbol{F}(t), e(t)), \tag{5.5.13}$$

where η^* is, as before, the response function for the entropy in equilibrium. In words this result tells us that *among all the processes ending* *with given values of the deformation gradient and the internal energy density the constant process has the greatest entropy.* This result is familiar to us from the theory of section 3.4.

From the maximal property of the entropy the thermostatic formulae

$$S^* = \frac{-\dfrac{\partial \eta^*}{\partial \boldsymbol{F}}}{\dfrac{\partial \eta^*}{\partial e}}, \qquad \theta^* = \frac{1}{\dfrac{\partial \eta^*}{\partial e}} \tag{5.5.14}$$

can be deduced using the same kind of argument as was used in section 5.4; these formulae too have been derived before in section 3.3 by a different route.

Because of (5.5.6), (5.5.8), (5.5.9) and (5.5.12) the inequality

$$\dot{\eta} \geqslant \frac{1}{\theta} \dot{e} - \frac{1}{\theta} \boldsymbol{S} \cdot \dot{\boldsymbol{F}} \tag{5.5.15}$$

holds in every process meeting the chain-rule conditions; in terms of the total heat supply h defined by equation (2.4.7) this is just the *local Clausius-Planck* inequality

$$\dot{\eta} \geqslant \frac{1}{\theta} h \tag{5.5.16}$$

which can be integrated to give[1]

$$\int_{t_0}^{t_1} \frac{1}{\theta(t)} h(t)\, dt \leqslant \eta(t_1) - \eta(t_0). \tag{5.5.17}$$

It follows from (5.5.15) that if $\dot{\boldsymbol{F}}=0$ and $\dot{e}=0$ then $\dot{\eta} \geqslant 0$, that is to say *the entropy cannot decrease if the deformation gradient and the internal energy density are held fixed.*

If it happens that the deformation gradient and the internal energy density are both constant at all times prior to t_0 then the entropy at that time is

$$\eta(t_0) = \eta^*(\boldsymbol{F}(t_0), e(t_0)).$$

However, according to (5.5.13) the entropy at a later time t_0 satisfies

$$\eta(t_1) \leqslant \eta^*(\boldsymbol{F}(t_1), e(t_1))$$

---

[1] Coleman has remarked (see Remark 28 of [8]) that the fact that, for the class of materials considered here, the Clausius-Duhem inequality implies the Clausius-Planck inequality justifies the assumptions of an earlier work of his [7] on the thermodynamics of incompressible viscoelastic fluids.

and so in this case the Clausius-Planck inequality implies that

$$\int_{t_0}^{t_1} \frac{1}{\theta(t)} h(t)\,dt \leqslant \eta^*(\boldsymbol{F}(t_1), e(t_1)) - \eta^*(\boldsymbol{F}(t_0), e(t_0)), \tag{5.5.18}$$

which is in agreement with (3.3.14). In particular if the path not only starts from equilibrium but is also closed in the sense that the deformation gradient and the internal energy return to their original values, that is if $\boldsymbol{F}(t_1)=\boldsymbol{F}(t_0)$ and $e(t_1)=e(t_0)$ then the *Clausius inequality*

$$\int_{t_0}^{t_1} \frac{1}{\theta(t)} h(t)\,dt \leqslant 0 \tag{5.5.19}$$

holds, again in agreement with the results of section 3.1.

It is easily checked that if the internal energy is used as an independent variable exactly the same conclusions can be drawn about the heat flux vector as were drawn in section 5.3: there is no piezo-caloric effect in equilibrium, the conductivity tensor in equilibrium is positive semi-definite and Fourier's law holds approximately near to equilibrium.

Now that we have the results of this section to hand it is possible, to a certain extent, to compare the theory based on the thermodynamic inequality and the theory based on the Clausius-Duhem inequality. Bearing in mind the special assumption made about the materials in both theories it is clear that for extensive classes of materials they agree in implying that the Clausius-Planck and the Clausius inequalities hold, that the entropy has the maximal property, that the equilibrium entropy determines the equilibrium stress and the equilibrium temperature through the classical formulae of thermostatics, that there is no piezo-caloric effect in equilibrium, that the conductivity tensor in equilibrium is positive semi-definite and that Fourier's law holds near to equilibrium.

Furthermore it is easy to see that for the materials considered in this section the validity of the Clausius-Duhem inequality in every process compatible with the balance laws for mass, momentum, moment of momentum and energy implies the validity of the thermodynamic inequality. This is so because in any process which is cyclic in the sense of section 2.2, namely $\boldsymbol{F}(\cdot)$ and $e(\cdot)$ are identically constant and $\boldsymbol{g}(t)$ vanishes identically at all times $t \leqslant t_0$ and $\boldsymbol{F}(t_1)=\boldsymbol{F}(t_0)$ and $e(t_1)=e(t_0)$, the Clausius-Duhem inequality

$$\dot{\eta} \geqslant -\frac{1}{\rho}\operatorname{div}\left(\frac{1}{\theta}\boldsymbol{q}\right) + \frac{1}{\theta} r$$

integrates to give

$$\int_{t_0}^{t_1} \left\{ -\frac{1}{\rho} \operatorname{div}\left(\frac{1}{\theta} \boldsymbol{q}\right) + \frac{1}{\theta} r \right\} dt \leqslant \eta(t_1) - \eta(t_0)$$

and, because of the maximal property of the entropy,

$$\begin{aligned} \eta(t_1) &= \check{\eta}(\boldsymbol{F}(t_1), e(t_1), \boldsymbol{F}^{t_1}(\cdot), e^{t_1}(\cdot)) \\ &\leqslant \eta^*(\boldsymbol{F}(t_1), e(t_1)) \\ &= \eta^*(\boldsymbol{F}(t_0), e(t_0)) \\ &= \eta(t_0) \end{aligned}$$

and so we recover the thermodynamic inequality

$$\int_{t_0}^{t_1} \left\{ -\frac{1}{\rho} \operatorname{div}\left(\frac{1}{\theta} \boldsymbol{q}\right) + \frac{1}{\theta} r \right\} dt \leqslant 0$$

which is none other than the starting point of the first theory.

The converse does not appear to be generally true; that is to say within the context of the first theory and with the entropy defined as it was in Chapter 3 it does not seem to follow that the Clausius-Duhem inequality necessarily holds. However, the Clausius-Duhem does hold in the first theory if the heat flux vector is given by a constitutive equation

$$\boldsymbol{q}(t) = \boldsymbol{q}(\boldsymbol{F}(t), e(t), \boldsymbol{g}(t)), \tag{5.5.20}$$

that is if it depends only on the present values of the deformation gradient, the internal energy and the temperature gradient. We require the stress and the temperature to be independent of the temperature gradient but they may depend on the histories of the deformation gradient and the internal energy:

$$\boldsymbol{T}(t) = \underset{-\infty}{\overset{t}{\overline{\boldsymbol{T}}}} (\boldsymbol{F}(\cdot), e(\cdot)), \tag{5.5.21}$$

$$\theta(t) = \underset{-\infty}{\overset{t}{\overline{\theta}}} (\boldsymbol{F}(\cdot), e(\cdot)). \tag{5.5.22}$$

Then, as we have seen, the entropy

$$\eta(t) = \underset{-\infty}{\overset{t}{\overline{\eta}}} (\boldsymbol{F}(\cdot), e(\cdot)) \tag{5.5.23}$$

satisfies the Clausius-Planck inequality

$$\dot{\eta} \geqslant \frac{1}{\theta} (\dot{e} - \boldsymbol{S} \cdot \dot{\boldsymbol{F}}) \tag{5.5.24}$$

and the heat flux vector satisfies

$$0 \geqslant \frac{1}{\rho\theta^2} \boldsymbol{g} \cdot \boldsymbol{q}.$$

Adding these two inequalities produces the inequality

$$\dot{\eta} \geqslant \frac{1}{\theta}(\dot{e} - \boldsymbol{S} \cdot \dot{\boldsymbol{F}}) + \frac{1}{\rho\theta^2} \boldsymbol{g} \cdot \boldsymbol{q}$$

which is equivalent to the Clausius-Duhem inequality. Loosely speaking, then, the two theories are equivalent for materials satisfying constitutive equations like (5.5.20), (5.5.21) and (5.5.22) but in general the theory based on the thermodynamic inequality is weaker (less restrictive) than the theory based on the Clausius-Duhem inequality.

It should be noted that even in the context of the theory based on the thermodynamic inequality it is quite possible, if the material behaves appropriately, to have the stress and the temperature determined by the entropy through the relations (5.5.8) and (5.5.9) not just in thermostatic equilibrium but away from equilibrium too. This will be the case if

I the response functionals (5.5.21) and (5.5.22) are such that the stress $\boldsymbol{T}(t)$ and the internal energy $e(t)$ are continuous functions of the time t on every path, and

II the entropy response functional (5.5.23) satisfies a chain-rule property[1] in the sense that there are functionals

$$\frac{\partial}{\partial \boldsymbol{F}} \overset{t}{\underset{-\infty}{\bar{\eta}}}(\boldsymbol{F}(\cdot), e(\cdot)), \qquad \frac{\partial}{\partial e} \overset{t}{\underset{-\infty}{\bar{\eta}}}(\boldsymbol{F}(\cdot), e(\cdot)), \qquad \delta \overset{t}{\underset{-\infty}{\bar{\eta}}}(\boldsymbol{F}(\cdot), e(\cdot)) \tag{5.5.25}$$

which are continuous functions of t on each path and which have the property that whenever the derivatives $\dot{\boldsymbol{F}}(t)$ and $\dot{e}(t)$ exist the derivative

$$\dot{\eta}(t) = \frac{d}{dt} \overset{t}{\underset{-\infty}{\bar{\eta}}}(\boldsymbol{F}(\cdot), e(\cdot))$$

exists and is given by

$$\begin{aligned} \frac{d}{dt} \overset{t}{\underset{-\infty}{\bar{\eta}}}(\boldsymbol{F}(\cdot), e(\cdot)) &= \frac{\partial}{\partial \boldsymbol{F}} \overset{t}{\underset{-\infty}{\bar{\eta}}}(\boldsymbol{F}(\cdot), e(\cdot)) \cdot \dot{\boldsymbol{F}}(t) \\ &\quad + \frac{\partial}{\partial e} \overset{t}{\underset{-\infty}{\bar{\eta}}}(\boldsymbol{F}(\cdot), e(\cdot)) \dot{e}(t) \\ &\quad + \delta \overset{t}{\underset{-\infty}{\bar{\eta}}}(\boldsymbol{F}(\cdot), e(\cdot)). \end{aligned} \tag{5.5.26}$$

[1] The chain-rule assumption *II* is, in essence, the one made by Gurtin [48]. Truesdell [67] calls it the strong chain-rule.

It is not difficult to show that the functionals (5.5.25) are unique if they exist. The specific example of a material which was considered in section 4.2 satisfies I and, as can be seen by differentiating the expression (4.2.59) for the entropy, it also satisfies II. The linearly viscous fluid does not satisfy I but it does satisfy II.

Suppose that we wish to establish that (5.5.8) and (5.5.9) do hold at a time t_0, say, where $(\boldsymbol{F}(s), e(s))$ is known for $s \leqslant t_0$ and assuming that I and II hold. As we have seen the Clausius-Planck inequality (5.5.24) must be satisfied wherever the derivatives $\dot{\eta}$, $\dot{\boldsymbol{F}}$ and $\dot{e}$ exist and, because of the chain-rule, the inequality can be put into the form

$$\left(\frac{\partial \overline{\eta}}{\partial \boldsymbol{F}}(t) + \frac{1}{\overline{\theta}(t)} \overline{\boldsymbol{S}}(t)\right) \cdot \dot{\boldsymbol{F}}(t) + \left(\frac{\partial \overline{\eta}}{\partial e}(t) - \frac{1}{\overline{\theta}(t)}\right) \dot{e}(t) + \delta \overline{\eta}(t) \geqslant 0 . \qquad (5.5.27)$$

If $\boldsymbol{A}$ is any tensor and if a is any scalar we can define a path $(\tilde{\boldsymbol{F}}(\cdot), \tilde{e}(\cdot))$ by $(\tilde{\boldsymbol{F}}(s), \tilde{e}(s)) = (\boldsymbol{F}(s), e(s))$ for $s \leqslant t_0$, by

$$(\tilde{\boldsymbol{F}}(s), \tilde{e}(s)) = (\boldsymbol{F}(t_0) + (s - t_0)\boldsymbol{A}, e(t_0) + (s - t_0)a)$$

for $t_0 \leqslant s \leqslant t_0 + \varepsilon$ and by $(\tilde{\boldsymbol{F}}(s), \tilde{e}(s)) = (\boldsymbol{F}(t_0) + \varepsilon \boldsymbol{A}, e(t_0) + \varepsilon a)$ for $s \geqslant t_0 + \varepsilon$ provided ε is sufficiently small. Applying (5.5.27) to this path at any time t in $t_0 < t < t_0 + \varepsilon$ gives

$$\left(\frac{\delta \overline{\eta}}{\partial \boldsymbol{F}}(t) + \frac{1}{\overline{\theta}(t)} \overline{\boldsymbol{S}}(t)\right) \cdot \boldsymbol{A} + \left(\frac{\partial \overline{\eta}}{\partial e}(t) - \frac{1}{\overline{\theta}(t)}\right) a + \delta \overline{\eta}(t) \geqslant 0 \qquad (5.5.28)$$

and if we now let $t \to t_0$ and use the continuity assumptions we are left with the conclusion that (5.5.28) holds at $t = t_0$ for any tensor $\boldsymbol{A}$ and any scalar a and this can be the case only if

$$\frac{\partial}{\partial \boldsymbol{F}} \overline{\eta}_{-\infty}^{t_0}(\boldsymbol{F}(\cdot), e(\cdot)) = \frac{-\overline{\boldsymbol{S}}_{-\infty}^{t_0}(\boldsymbol{F}(\cdot), e(\cdot))}{\overline{\theta}_{-\infty}^{t_0}(\boldsymbol{F}(\cdot), e(\cdot))},$$

$$\frac{\partial}{\partial e} \overline{\eta}_{-\infty}^{t_0}(\boldsymbol{F}(\cdot), e(\cdot)) = \frac{1}{\overline{\theta}_{-\infty}^{t_0}(\boldsymbol{F}(\cdot), e(\cdot))},$$

and

$$\delta \overline{\eta}_{-\infty}^{t_0}(\boldsymbol{F}(\cdot), e(\cdot)) \geqslant 0 .$$

In other words the formulae (5.5.8) and (5.5.9) do hold and so does the internal dissipation inequality (5.5.12).

CHAPTER 6

Thermodynamic Restrictions on Isothermal Linear Viscoelasticity

6.1 Compatibility with Thermodynamics. Dissipative Relaxation Functions

In the linearised theory of isothermal viscoelasticity the stress tensor is determined by the history of the infinitesimal strain tensor through the hereditary law[1]

$$\boldsymbol{T}(t)=\boldsymbol{T}(t,\boldsymbol{E}(\cdot))=\mathscr{G}(0)\,\boldsymbol{E}(t)+\int_0^{+\infty}\dot{\mathscr{G}}(s)\,\boldsymbol{E}(t-s)\,ds. \tag{6.1.1}$$

Here $\mathscr{G}(\cdot)$ is the relaxation function of the material; its values are fourth order tensor. In the suffix notation (6.1.1) becomes

$$T_{ij}(t)=\sum_{k,l}\left\{\mathscr{G}_{ijkl}(0)\,E_{kl}(t)+\int_0^{+\infty}\dot{\mathscr{G}}_{ijkl}(s)\,E_{kl}(t-s)\,ds\right\}. \tag{6.1.2}$$

Because the tensors $\boldsymbol{T}$ and $\boldsymbol{E}$ are symmetric it can be assumed that $\mathscr{G}(\cdot)$ satisfies the conditions

$$\mathscr{G}_{ijkl}(\cdot)=\mathscr{G}_{jikl}(\cdot)=\mathscr{G}_{ijlk}(\cdot) \tag{6.1.3}$$

which imply that at most 36 of the 81 functions $\mathscr{G}_{ijkl}(\cdot)$ can be independent.

If the strain suffers a jump discontinuity of amount

$$[\boldsymbol{E}(t)]=\boldsymbol{E}(t+)-\boldsymbol{E}(t-) \tag{6.1.4}$$

at time t then the corresponding jump in the stress is

$$[\boldsymbol{T}(t)]=\mathscr{G}(0)\,[\boldsymbol{E}(t)] \tag{6.1.5}$$

---

[1] The reader is referred to the article [50] of Gurtin and Sternberg for a comprehensive account of linear viscoelasticity. The status of the linearised theory as an approximation to the general theory of simple materials with fading memory is clarified by the work of Coleman and Noll [26] and of Coleman [9].

and, for that reason, it is usual to call the fourth order tensor $\mathscr{G}(0)$ the *instantaneous elastic modulus* of the material. It will be assumed throughout this Chapter that $\mathscr{G}(\cdot)$ has a continuous derivative $\dot{\mathscr{G}}(\cdot)$ and that

$$\int_0^{+\infty} \|\dot{\mathscr{G}}(t)\|\, dt < +\infty, \quad \int_0^{+\infty} t\|\dot{\mathscr{G}}(t)\|\, dt < +\infty. \tag{6.1.6}$$

These assumptions ensure that the *equilibrium elastic modulus*

$$\mathscr{G}(+\infty) = \lim_{t\to+\infty} \mathscr{G}(t)$$

exists[1]. The stress corresponding to the constant strain history $\boldsymbol{E}(\cdot) \equiv \boldsymbol{A}$ is

$$\boldsymbol{T}^*(\boldsymbol{A}) = \mathscr{G}(+\infty)\boldsymbol{A}, \tag{6.1.7}$$

which accounts for the name given to $\mathscr{G}(+\infty)$.

Not every relaxation function meeting the requirements (6.1.6) behaves in a physically realistic way and one way of restricting the behaviour of relaxation functions is by making appropriate assumptions about the material which have a thermodynamic character. These assumptions generally take the form of statements about the work done on certain strain paths; in this Chapter we will investigate the restrictions imposed by various statements of this kind and establish some connections between them. Before we do this however we shall need to make a few definitions. It will be recalled that the *scalar product* of the symmetric tensors $\boldsymbol{M}, \boldsymbol{N}$ was defined in section 1.1 to be $\boldsymbol{M}\cdot\boldsymbol{N} = \operatorname{trace} \boldsymbol{M}\boldsymbol{N}$ and that the *norm* of $\boldsymbol{M}$ was defined to be $|\boldsymbol{M}| = (\boldsymbol{M}\cdot\boldsymbol{M})^{\frac{1}{2}}$. By the *transpose* of the fourth order tensor $\mathscr{P}$ we shall mean the tensor $\mathscr{P}^T$ defined by the condition $\boldsymbol{M}\cdot\mathscr{P}\boldsymbol{N} = \boldsymbol{N}\cdot\mathscr{P}^T\boldsymbol{M}$, holding for all symmetric tensors $\boldsymbol{M}$ and $\boldsymbol{N}$. We say that $\mathscr{P}$ is *symmetric* if $\mathscr{P}^T = \mathscr{P}$ and *skew-symmetric* if $\mathscr{P}^T = -\mathscr{P}$; the symmetry of $\mathscr{P}$ is equivalent to the conditions $\mathscr{P}_{ijkl} = \mathscr{P}_{klij}$ on its cartesian components. We say too that $\mathscr{P}$ is *positive semi-definite* if $\boldsymbol{M}\cdot\mathscr{P}\boldsymbol{M} \geqslant 0$ for every symmetric $\boldsymbol{M}$, that is if $\sum_{i,j,k,l} \mathscr{P}_{ijkl} M_{ij} M_{kl} \geqslant 0$, that $\mathscr{P}$ is *positive definite* if $\boldsymbol{M}\cdot\mathscr{P}\boldsymbol{M} > 0$ for every $\boldsymbol{M} \neq \boldsymbol{0}$ and that $\mathscr{P}$

[1] Because the first of the integrals in (6.1.6) is finite we know that corresponding to any $\varepsilon > 0$ there is a number $t_0(\varepsilon)$, depending on ε, such that

$$\int_{t_1}^{t_2} \|\dot{\mathscr{G}}(t)\|\, dt < \varepsilon$$

whenever $t_0(\varepsilon) < t_1 < t_2$. Since

$$\|\mathscr{G}(t_1) - \mathscr{G}(t_2)\| \leqslant \int_{t_1}^{t_2} \|\dot{\mathscr{G}}(t)\|\, dt < \varepsilon$$

this means that $\lim_{t\to+\infty} \mathscr{G}(t)$ exists.

is *negative semi-definite* if $-\mathscr{P}$ is positive semi-definite. If the difference $\mathscr{P}_1 - \mathscr{P}_2$ of two tensors $\mathscr{P}_1, \mathscr{P}_2$ is positive semi-definite we shall write $\mathscr{P}_1 \geqslant \mathscr{P}_2$; clearly the inequalities $\mathscr{P}_1 \geqslant \mathscr{P}_2$ and $\mathscr{P}_2 \geqslant \mathscr{P}_3$ together imply the inequality $\mathscr{P}_1 \geqslant \mathscr{P}_3$. The *norm* $\|\mathscr{P}\|$ of the tensor $\mathscr{P}$ is the least upper bound of the numbers $|\mathscr{P}\boldsymbol{M}|$ obtained as $\boldsymbol{M}$ ranges over all symmetric tensors with $|\boldsymbol{M}| = 1$.

In his paper [9] Coleman discussed the foundations of linear viscoelasticity, viewed as an approximation to a general theory of simple materials with fading memory, and then he turned to investigating thermodynamic restrictions on relaxation functions. He was able to show as a consequence of the minimal property (5.4.13) of the free energy that the difference $\mathscr{G}(0) - \mathscr{G}(+\infty)$ between the instantaneous and the equilibrium elastic moduli must be positive semi-definite and symmetric. Since it follows from the relation (5.4.20) for the equilibrium stress that $\mathscr{G}(+\infty)$ is symmetric Coleman's work implies that the elastic moduli $\mathscr{G}(0)$, $\mathscr{G}(+\infty)$ are both symmetric and that, in our notation,

$$\mathscr{G}(0) \geqslant \mathscr{G}(+\infty). \tag{6.1.8}$$

In section 5.4 it was shown, for the class of materials considered there, that the inequality (5.4.16) must hold; it asserts that the work done around any closed isothermal strain path starting from equilibrium cannot be negative. In the present context this means that

$$\int_{-\infty}^{+\infty} \boldsymbol{T}(t, \boldsymbol{E}(\cdot)) \cdot \dot{\boldsymbol{E}}(t)\, dt \geqslant 0 \tag{6.1.9}$$

whenever $\boldsymbol{E}(\cdot)$ is a closed strain path in the sense that there is a symmetric tensor $\boldsymbol{A}$ and times t_0, t_1 with $\boldsymbol{E}(t) = \boldsymbol{A}$ for every $t \leqslant t_0$ and for every $t \geqslant t_1$. We shall say that the relaxation function $\mathscr{G}(\cdot)$ is *compatible with thermodynamics* if this requirement is satisfied. Clearly compatibility with thermodynamics severely restricts the behaviour of $\mathscr{G}(\cdot)$.

A different restriction on relaxation functions, which also has a thermodynamic character but which does not follow from the Clausius-Duhem inequality without an additional assumption is the one proposed by König and Meixner [55] in a study of one-dimensional constitutive relations. This restriction is the requirement that

$$\int_{-\infty}^{u} \boldsymbol{T}(t, \boldsymbol{E}(\cdot)) \cdot \dot{\boldsymbol{E}}(t)\, dt \geqslant 0 \tag{6.1.10}$$

for every time u and for every path starting from the virgin state of zero strain in the sense that $\boldsymbol{E}(-\infty) = \boldsymbol{O}$, that is to say there is a time t_0 such that $\boldsymbol{E}(t) \equiv \boldsymbol{O}$ for every $t \leqslant t_0$. The relaxation function is said to be *dissipative* if this requirement holds.

The implications of dissipativity for linear viscoelastic materials were first investigated by Shu and Onat [65] and by Gurtin and Herrera [49]. Shu and Onat were able to show that if $\mathscr{G}(\cdot)$ is dissipative then the instantaneous elastic modulus $\mathscr{G}(0)$ is symmetric and Gurtin and Herrera showed, by arguments of a different kind, that

I If $\mathscr{G}(\cdot)$ is dissipative then the elastic moduli $\mathscr{G}(0)$ and $\mathscr{G}(+\infty)$ are both symmetric and positive semi-definite and

$$\mathscr{G}(0) \geqslant \pm \mathscr{G}(s) \tag{6.1.11}$$

for every $s \geqslant 0$. In particular, the inequality (6.1.8) *can be strengthened to*

$$\mathscr{G}(0) \geqslant \mathscr{G}(+\infty) \geqslant \mathbf{0} \tag{6.1.12}$$

and

$$-\dot{\mathscr{G}}(0) \geqslant \mathbf{0}. \tag{6.1.13}$$

The statement *I* will be proved below[1]. We shall also establish the following connection between dissipative relaxation functions and those which are compatible with thermodynamics:

II (i) *$\mathscr{G}(\cdot)$ is dissipative if and only if it is compatible with thermodynamics and the equilibrium elastic modulas $\mathscr{G}(+\infty)$ is positive semidefinite,* (ii) *$\mathscr{G}(\cdot)$ is compatible with thermodynamics if and only if the relaxation function $\mathscr{G}(\cdot)-\mathscr{G}(+\infty)$ is dissipative and the equilibrium elastic modulus $\mathscr{G}(+\infty)$ is symmetric.*

Once *I* and *II* have been proved it follows immediately that

III If $\mathscr{G}(\cdot)$ is compatible with thermodynamics, and in particular if $\mathscr{G}(\cdot)$ is dissipative, then the elastic moduli $\mathscr{G}(0)$ and $\mathscr{G}(+\infty)$ are both symmetric and

$$\mathscr{G}(0)-\mathscr{G}(+\infty) \geqslant +(\mathscr{G}(s)-\mathscr{G}(+\infty)) \tag{6.1.14}$$

for every $s \geqslant 0$. Furthermore the inequalities (6.1.8) *and* (6.1.13) *hold.*

Inequalities of the kind (6.1.12) and (6.1.13) have important consequenses for wave propagation—the positive definiteness of $\mathscr{G}(0)$ and the negative definiteness of $\dot{\mathscr{G}}(0)$ together ensure the decay of plane shock and acceleration waves[2].

In order to prove *I, II* and *III*[3] we shall need to prove a number of preliminary results. As before we say that a function $\boldsymbol{E}(\cdot)$ whose values are symmetric tensors is a strain *path* if there are tensors $\boldsymbol{A}_0$, $\boldsymbol{A}_1$ and times t_0, t_1 with $\boldsymbol{E}(t) \equiv \boldsymbol{A}_0$ for every $t \leqslant t_0$, with $\boldsymbol{E}(t) \equiv \boldsymbol{A}_1$ for every $t \geqslant t_1$ and

[1] The argument used to establish (6.1.11) differs from that of Gurtin and Herrera in that it makes no direct appeal to known properties of functions of positive type.

[2] See Coleman, Gurtin and Herrera [17].

[3] The proof is essentially that of [38].

if $\boldsymbol{E}(\cdot)$ is continuous and has a piecewise continuous derivative $\dot{\boldsymbol{E}}(\cdot)$ defined everywhere except possibly at a finite number of values of t. We write $\boldsymbol{E}(-\infty)$ for the *initial value* $\boldsymbol{A}_0$ and $\boldsymbol{E}(+\infty)$ for the *final value* $\boldsymbol{A}_1$. The path is said to be *closed* if $\boldsymbol{E}(-\infty)=\boldsymbol{E}(+\infty)$ and it is said to *start from the virgin state* if $\boldsymbol{E}(-\infty)=\boldsymbol{O}$. If $\boldsymbol{E}(\cdot)$ is any path and if τ is any fixed time we can introduce, as before, a new path $\boldsymbol{E}(\cdot,\alpha)$ with

$$\boldsymbol{E}(t,\alpha)=\boldsymbol{E}(\tau+\alpha(t-\tau)). \tag{6.1.15}$$

This path is a *retardation* of the path $\boldsymbol{E}(\cdot)$ if $0<\alpha<1$ and an *acceleration* of it if $\alpha>1$. The *work done* on the path $\boldsymbol{E}(\cdot)$ is the integral

$$w(\boldsymbol{E}(\cdot))=\int_{-\infty}^{+\infty}\boldsymbol{T}(t,\boldsymbol{E}(\cdot))\cdot\dot{\boldsymbol{E}}(t)\,dt, \tag{6.1.16}$$

which is well-defined because the integrand vanishes outside some interval of finite length. The relaxation function is *compatible with thermodynamics* if

$$w(\boldsymbol{E}(\cdot))\geqslant 0 \tag{6.1.17}$$

for every closed path $\boldsymbol{E}(\cdot)$ and it is *dissipative* if (6.1.17) holds for every path starting from the virgin state.

We begin by proving two results describing the behaviour of the material under extreme retardation and extreme acceleration of an arbitrary path $\boldsymbol{E}(\cdot)$:

$$\lim_{\alpha\to 0} w(\boldsymbol{E}(\cdot,\alpha))=\int_{-\infty}^{+\infty}\dot{\boldsymbol{E}}(t)\cdot\mathscr{G}(+\infty)\boldsymbol{E}(t)\,dt, \tag{6.1.18}$$

$$\begin{aligned}\lim_{\alpha\to +\infty} w(\boldsymbol{E}(\cdot,\alpha))=&\int_{-\infty}^{+\infty}\dot{\boldsymbol{E}}(t)\cdot\mathscr{G}(0)\boldsymbol{E}(t)\,dt\\ &+(\boldsymbol{E}(+\infty)-\boldsymbol{E}(-\infty))\cdot(\mathscr{G}(0)-\mathscr{G}(+\infty))\boldsymbol{E}(-\infty).\end{aligned} \tag{6.1.19}$$

These results shed further light on the physical meaning of the two elastic moduli $\mathscr{G}(0)$, $\mathscr{G}(+\infty)$.

According to (6.1.15) and (6.1.16)

$$\begin{aligned} w(\boldsymbol{E}(\cdot,\alpha)) &= \int_{-\infty}^{+\infty}\boldsymbol{T}(t,\boldsymbol{E}(\cdot,\alpha))\cdot\dot{\boldsymbol{E}}(t,\alpha)\,dt\\ &= \int_{-\infty}^{+\infty}\alpha\dot{\boldsymbol{E}}(\tau+\alpha(t-\tau))\cdot\Big\{\mathscr{G}(0)\boldsymbol{E}(\tau+\alpha(t-\tau))\\ &\qquad +\int_0^{+\infty}\dot{\mathscr{G}}(s)\boldsymbol{E}(\tau+\alpha(t-s-\tau))\,ds\Big\}\,dt\end{aligned}$$

and on making the change of variable $u=\tau+\alpha(t-\tau)$ we deduce that

$$w(\boldsymbol{E}(\cdot,\alpha))=\int_{-\infty}^{+\infty}\dot{\boldsymbol{E}}(u)\cdot\left\{\mathscr{G}(0)\boldsymbol{E}(u)+\int_{0}^{+\infty}\dot{\mathscr{G}}(s)\boldsymbol{E}(u-\alpha s)\,ds\right\}du \tag{6.1.20}$$

$$=\int_{-\infty}^{+\infty}\dot{\boldsymbol{E}}(u)\cdot\mathscr{G}(+\infty)\,\boldsymbol{E}(u)\,du$$

$$+\int_{-\infty}^{+\infty}\int_{0}^{+\infty}\dot{\boldsymbol{E}}(u)\cdot\dot{\mathscr{G}}(s)(\boldsymbol{E}(u-\alpha s)-\boldsymbol{E}(u))\,ds\,du. \tag{6.1.21}$$

However if c_1 is an upper bound for the derivative $|\dot{\boldsymbol{E}}(\cdot)|$ the mean value theorem implies that

$$|\boldsymbol{E}(u-\alpha s)-\boldsymbol{E}(u)|\leqslant c_1\,\alpha\,s \tag{6.1.22}$$

for every u and every $s\geqslant 0$. Since $\dot{\boldsymbol{E}}(\cdot)$ vanishes outside a finite interval and since the integral

$$\int_{0}^{+\infty} s\|\dot{\mathscr{G}}(s)\|\,ds$$

has been assumed to be finite it follows from (6.1.21) and (6.1.22) that

$$\left|\int_{-\infty}^{+\infty}\int_{0}^{+\infty}\dot{\boldsymbol{E}}(u)\cdot\mathscr{G}(s)(\boldsymbol{E}(u-\alpha s)-\boldsymbol{E}(u))\,ds\,du\right|=O(\alpha)$$

as $\alpha\to 0$ and the result (6.1.18) follows.

To prove (6.1.19) observe that because of (6.1.20)

$$w(\boldsymbol{E}(\cdot,\alpha))=\int_{-\infty}^{+\infty}\dot{\boldsymbol{E}}(u)\cdot\mathscr{G}(0)\,\boldsymbol{E}(u)\,du$$

$$+(\boldsymbol{E}(+\infty)-\boldsymbol{E}(-\infty))\cdot(\mathscr{G}(0)-\mathscr{G}(\infty))\,\boldsymbol{E}(-\infty)$$

$$+\int_{-\infty}^{+\infty}\int_{0}^{+\infty}\dot{\boldsymbol{E}}(u)\cdot\dot{\mathscr{G}}(s)(\boldsymbol{E}(u-\alpha s)-\boldsymbol{E}(-\infty)\,ds\,du. \tag{6.1.23}$$

If we choose t_0, t_1 so that $\boldsymbol{E}(t)=\boldsymbol{E}(-\infty)$ for every $t\leqslant t_0$ and $\boldsymbol{E}(t)=\boldsymbol{E}(+\infty)$ for every $t\geqslant t_1$ we have

$$\left|\int_{-\infty}^{+\infty}\int_{0}^{+\infty}\dot{\boldsymbol{E}}(u)\cdot\dot{\mathscr{G}}(s)(\boldsymbol{E}(u-\alpha s)-\boldsymbol{E}(-\infty))\,ds\,du\right|$$

$$=\left|\int_{t_0}^{t_1}\int_{0}^{(u-t_0)/\alpha}\dot{\boldsymbol{E}}(u)\cdot\dot{\mathscr{G}}(s)(\boldsymbol{E}(u-\alpha s)-\boldsymbol{E}(-\infty))\,ds\,du\right.$$

$$\leqslant c_2\int_{t_0}^{t_1}\int_{0}^{(u-t_0)/\alpha}ds\,du=\frac{c_2}{2\alpha}(t_1-t_0)^2$$

for some appropriate constant c_2 and the required result (6.1.19) now follows on letting $\alpha\to+\infty$ in (6.1.23).

We turn to proving the statement *I*. To do this let $\boldsymbol{E}(\cdot)$ be any path starting from the virgin state. Then if $\boldsymbol{E}(\cdot,\alpha)$ is defined as in (6.1.15) it too is a path starting from the virgin state and if the relaxation function $\mathscr{G}(\cdot)$ is dissipative we must have

$$w(\boldsymbol{E}(\cdot,\alpha)) \geqslant 0.$$

Consequently

$$\lim_{\alpha \to +\infty} w(\boldsymbol{E}(\cdot,\alpha)) \geqslant 0$$

and

$$\lim_{\alpha \to 0} w(\boldsymbol{E}(\cdot,\alpha)) \geqslant 0$$

and the expressions (6.1.19) and (6.1.18) for these limits tell us, since $\boldsymbol{E}(-\infty)=\boldsymbol{O}$, that

$$\int_{-\infty}^{+\infty} \dot{\boldsymbol{E}}(t)\cdot\mathscr{G}(0)\boldsymbol{E}(t)\,dt \geqslant 0 \tag{6.1.24}$$

and

$$\int_{-\infty}^{+\infty} \dot{\boldsymbol{E}}(t)\cdot\mathscr{G}(+\infty)\boldsymbol{E}(t)\,dt \geqslant 0 \tag{6.1.25}$$

for any such path. In particular if we take the path to be closed we deduce, as in section 1.2, that the moduli $\mathscr{G}(0)$ and $\mathscr{G}(+\infty)$ are both symmetric. Furthermore if $\boldsymbol{A}$ is any symmetric tensor we can choose a path $\boldsymbol{E}(\cdot)$ with $\boldsymbol{E}(-\infty)=\boldsymbol{O}$ and $\boldsymbol{E}(+\infty)=\boldsymbol{A}$ and then, because $\mathscr{G}(0)$ and $\mathscr{G}(+\infty)$ are symmetric, the integrations in (6.1.24) and (6.1.25) can be performed to give the inequalities

$$\tfrac{1}{2}\boldsymbol{A}\cdot\mathscr{G}(0)\boldsymbol{A} \geqslant 0, \quad \tfrac{1}{2}\boldsymbol{A}\cdot\mathscr{G}(+\infty)\boldsymbol{A} \geqslant 0,$$

that is to say $\mathscr{G}(0)$ and $\mathscr{G}(+\infty)$ are both positive semi-definite—which proves the first part of *I*.

To prove the second part we proceed as follows: observe that if the path $\boldsymbol{E}(\cdot)$ starts from the virgin state an integration by parts shows that (6.1.1) can be written as

$$\boldsymbol{T}(t,\boldsymbol{E}(\cdot)) = \int_{-\infty}^{t} \mathscr{G}(t-u)\dot{\boldsymbol{E}}(u)\,du \tag{6.1.26}$$

and the work done as

$$w(\boldsymbol{E}(\cdot)) = \int_{-\infty}^{+\infty}\int_{-\infty}^{t} \dot{\boldsymbol{E}}(t)\cdot\mathscr{G}(t-u)\dot{\boldsymbol{E}}(u)\,du\,dt. \tag{6.1.27}$$

If $s>0$ is any fixed positive number, if $\boldsymbol{A}, \boldsymbol{B}$ are any symmetric tensors and if $0<\varepsilon<u$ we can define a piecewise linear path starting from the virgin state by setting $\boldsymbol{E}(t)=\boldsymbol{O}$ for $t\leqslant 0$, $\boldsymbol{E}(t)=(1/\varepsilon)t\boldsymbol{A}$ for $0\leqslant t\leqslant\varepsilon$, $\boldsymbol{E}(t)=\boldsymbol{A}$ for $\varepsilon\leqslant t\leqslant s$, $\boldsymbol{E}(t)=\boldsymbol{A}+(1/\varepsilon)(t-s)\boldsymbol{B}$ for $s\leqslant t\leqslant s+\varepsilon$ and

$\boldsymbol{E}(t)=\boldsymbol{A}+\boldsymbol{B}$ for $t \geqslant s+\varepsilon$ (see Fig. 7). If we compute the work done on this path using (6.1.27) the inequality (6.1.17) tells us that

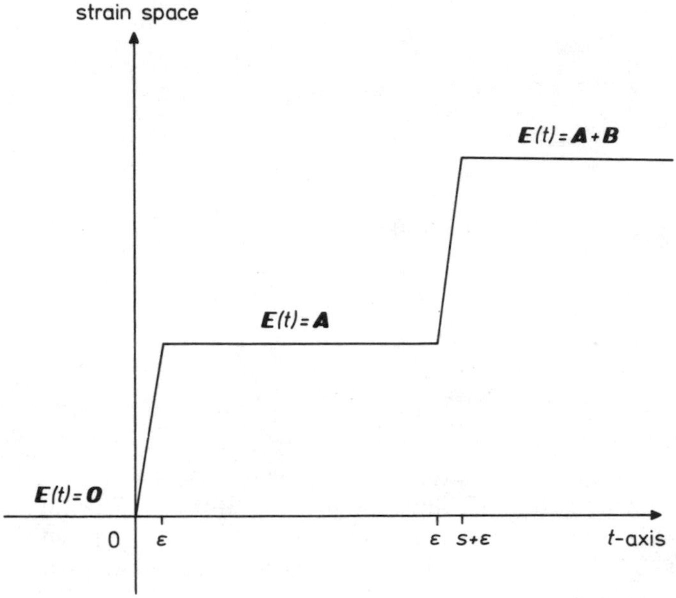

Fig. 7

$$\frac{1}{\varepsilon^2}\int_0^{\varepsilon}\int_0^{t} \boldsymbol{A}\cdot\mathscr{G}(t-u)\boldsymbol{A}\,du\,dt+\frac{1}{\varepsilon^2}\int_s^{s+\varepsilon}\int_0^{\varepsilon} \boldsymbol{B}\cdot\mathscr{G}(t-u)\boldsymbol{A}\,du\,dt$$
$$+\frac{1}{\varepsilon^2}\int_s^{s+\varepsilon}\int_s^{t} \boldsymbol{B}\cdot\mathscr{G}(t-u)\boldsymbol{B}\,du\,dt \geqslant 0\,.$$

On letting $\varepsilon \to 0$ we obtain the inequality

$$\tfrac{1}{2}\boldsymbol{A}\cdot\mathscr{G}(0)\boldsymbol{A}+\boldsymbol{B}\cdot\mathscr{G}(s)\boldsymbol{A}+\tfrac{1}{2}\boldsymbol{B}\cdot\mathscr{G}(0)\boldsymbol{B} \geqslant 0\,. \tag{6.1.28}$$

If we take $\boldsymbol{B}=-\boldsymbol{A}$ we find that

$$\mathscr{G}(0) \geqslant \mathscr{G}(s) \tag{6.1.29}$$

and if we take $\boldsymbol{B}=\boldsymbol{A}$ we find that

$$\mathscr{G}(0) \geqslant -\mathscr{G}(s)\,. \tag{6.1.30}$$

In other words the required inequality (6.1.11) holds. If we let $s \to +\infty$ in (6.1.28) and remember that $\mathscr{G}(+\infty)$ has already been shown to be positive semi-definite we deduce (6.1.12). The last assertion of *I*, namely (6.1.13), is an immediate consequence of (6.1.29).

Next we turn to proving *II*. If $\boldsymbol{E}(\cdot)$ is any path we can associate with it a path $\hat{\boldsymbol{E}}(\cdot)$ starting from the virgin state by setting

$$\boldsymbol{E}(t) = \boldsymbol{E}(-\infty) + \hat{\boldsymbol{E}}(t), \qquad -\infty < t < +\infty. \tag{6.1.31}$$

According to (6.1.1) the stress at time t on the two paths is related by

$$\boldsymbol{T}(t, \boldsymbol{E}(\cdot)) = \boldsymbol{T}(t, \hat{\boldsymbol{E}}(\cdot)) + \mathscr{G}(+\infty)\boldsymbol{E}(-\infty) \tag{6.1.32}$$

and the work done on the two paths by

$$w(\boldsymbol{E}(\cdot)) = w(\hat{\boldsymbol{E}}(\cdot)) + (\boldsymbol{E}(+\infty) - \boldsymbol{E}(-\infty)) \cdot \mathscr{G}(+\infty)\boldsymbol{E}(-\infty). \tag{6.1.33}$$

If $\mathscr{G}(\cdot)$ is dissipative we must have

$$w(\hat{\boldsymbol{E}}(\cdot)) \geqslant 0$$

and hence

$$w(\boldsymbol{E}(\cdot)) \geqslant (\boldsymbol{E}(+\infty) - \boldsymbol{E}(-\infty)) \cdot \mathscr{G}(+\infty)\boldsymbol{E}(-\infty).$$

In particular, if $\boldsymbol{E}(\cdot)$ is any closed path we have $\boldsymbol{E}(-\infty) = \boldsymbol{E}(+\infty)$ and thus

$$w(\boldsymbol{E}(\cdot)) \geqslant 0,$$

that is to say $\mathscr{G}(\cdot)$ is compatible with thermodynamics. Since we have shown already that $\mathscr{G}(+\infty)$ is positive semi-definite whenever $\mathscr{G}(\cdot)$ is dissipative we have now demonstrated the truth of the 'only if' part of *II* (i).

To prove the 'if' part of *II* (i) let us take it that $\mathscr{G}(\cdot)$ is compatible with thermodynamics. We begin by showing that $\mathscr{G}(+\infty)$ must be symmetric. This is easily seen to be the case because if $\boldsymbol{E}_1(\cdot)$ is any closed path, its retardation $\boldsymbol{E}_1(\cdot, \alpha)$ is also closed and hence

$$w(\boldsymbol{E}_1(\cdot, \alpha)) \geqslant 0.$$

On letting $\alpha \to 0$ and using (6.1.18) we find that

$$\int_{-\infty}^{+\infty} \dot{\boldsymbol{E}}_1(t) \cdot \mathscr{G}(+\infty)\boldsymbol{E}_1(t)\,dt \geqslant 0$$

whenever $\boldsymbol{E}_1(\cdot)$ is closed and hence $\mathscr{G}(+\infty)$ is symmetric. Next, still on the assumption that $\mathscr{G}(\cdot)$ is compatible with thermodynamics, we shall show that if $\boldsymbol{E}(\cdot)$ starts from the virgin state then

$$w(\boldsymbol{E}(\cdot)) \geqslant \tfrac{1}{2}\boldsymbol{E}(+\infty) \cdot \mathscr{G}(+\infty)\boldsymbol{E}(+\infty). \tag{6.1.34}$$

Once we have proved this inequality we can deduce that if $\mathscr{G}(+\infty)$ is positive semi-definite then

$$w(\boldsymbol{E}(\cdot)) \geqslant 0,$$

that is to say $\mathscr{G}(\cdot)$ is dissipative. The proof of (6.1.33) requires arguments of the kind used in section 4.2. Let us choose t_0, t_1 in such a way that $\boldsymbol{E}(t) = \boldsymbol{O}$ for $t \leqslant t_0$ and $\boldsymbol{E}(t) \equiv \boldsymbol{E}(+\infty)$ for $t \geqslant t_1$. Then if $\boldsymbol{H}(\cdot)$ is any strain path with $\boldsymbol{H}(\cdot) = \boldsymbol{E}(+\infty)$ for $t \leqslant 0$ and $\boldsymbol{H}(t) = \boldsymbol{O}$ for $t \geqslant 1$ we can construct a double sequence of paths $\boldsymbol{E}_{\alpha\lambda}(t)$, for $0 < \alpha < 1$ and $\lambda > 0$, by setting $\boldsymbol{E}_{\alpha\lambda}(t) = \boldsymbol{E}(t+\lambda)$ for $t \leqslant t_1 - \lambda$ and $\boldsymbol{E}_{\alpha\lambda}(t) = \boldsymbol{H}(t,\alpha)$ for $t \geqslant t_1 - \lambda$ where $\boldsymbol{H}(t,\alpha) = \boldsymbol{H}(\alpha t)$ is a retardation of $\boldsymbol{H}(\cdot)$. Each of the paths $\boldsymbol{E}_{\alpha\lambda}(\cdot)$ is closed and thus, because $\mathscr{G}(\cdot)$ is compatible with thermodynamics,

$$\begin{aligned} 0 \leqslant w(\boldsymbol{E}_{\alpha\lambda}(\cdot)) &= \int_{t_0-\lambda}^{1/\alpha} \boldsymbol{T}(t, \boldsymbol{E}_{\alpha\lambda}(\cdot)) \cdot \dot{\boldsymbol{E}}_{\alpha\lambda}(t)\, dt \\ &= \left(\int_{t_0-\lambda}^{t_1-\lambda} + \int_0^{1/\alpha} \right) \boldsymbol{T}(t, \boldsymbol{E}_{\alpha\lambda}(\cdot)) \cdot \dot{\boldsymbol{E}}_{\alpha\lambda}(t)\, dt \\ &= \int_{t_0}^{t_1} \boldsymbol{T}(t, \boldsymbol{E}(\cdot)) \cdot \dot{\boldsymbol{E}}(t)\, dt + \int_0^{1/\alpha} \boldsymbol{T}(t, \boldsymbol{E}_{\alpha\lambda}(\cdot)) \cdot \dot{\boldsymbol{H}}(t,\alpha)\, dt, \end{aligned}$$

that is to say

$$0 \leqslant w(\boldsymbol{E}(\cdot)) + \int_0^{1/\alpha} \boldsymbol{T}(t, \boldsymbol{E}_{\alpha\lambda}(\cdot)) \cdot \dot{\boldsymbol{H}}(t,\alpha)\, dt. \tag{6.1.35}$$

But, for $0 \leqslant t \leqslant 1/\alpha$,

$$\begin{aligned} \boldsymbol{T}(t, \boldsymbol{E}_{\alpha\lambda}(\cdot)) &= \mathscr{G}(0)\boldsymbol{H}(t,\alpha) + \int_0^{+\infty} \dot{\mathscr{G}}(s) \boldsymbol{E}_{\alpha\lambda}(t-s)\, ds \\ &= \mathscr{G}(0)\boldsymbol{H}(t,\alpha) + \int_0^{t-t_1+\lambda} \dot{\mathscr{G}}(s) \boldsymbol{H}(t-s,\alpha)\, ds \\ &\quad + \int_{t-t_1+\lambda}^{+\infty} \dot{\mathscr{G}}(s) \boldsymbol{E}(t-s+\lambda)\, ds \end{aligned}$$

and hence

$$\boldsymbol{T}(t, \boldsymbol{E}_{\alpha\lambda}(\cdot)) - \boldsymbol{T}(t, \boldsymbol{H}(\cdot,\alpha)) = \int_{t-t_1+\lambda}^{+\infty} \dot{\mathscr{G}}(s)(\boldsymbol{E}(t-s+\lambda) - \boldsymbol{H}(t-s,\alpha))\, ds \tag{6.1.36}$$

and it is now a straightforward matter to deduce from (6.1.36) the estimate

$$\left| \int_0^{1/\alpha} \boldsymbol{T}(t, \boldsymbol{E}_{\alpha\lambda}(\cdot)) \cdot \dot{\boldsymbol{H}}(t,\alpha)\, dt - \int_0^{1/\alpha} \boldsymbol{T}(t, \boldsymbol{H}(\cdot,\alpha)) \cdot \dot{\boldsymbol{H}}(t,\alpha)\, dt \right| \leqslant c_3 \int_{\lambda-t_1}^{+\infty} \|\dot{\mathscr{G}}(s)\|\, ds \tag{6.1.37}$$

for some constant c_3. Since

$$\int_{\lambda-t_1}^{+\infty} \|\dot{\mathscr{G}}(s)\|\, ds \to 0$$

as $\lambda \to +\infty$ it follows from (6.1.35) and (6.1.36) that on letting $\lambda \to +\infty$ in (6.1.34) we are left with the inequality

$$0 \leqslant w(\boldsymbol{E}(\cdot)) + w(\boldsymbol{H}(\cdot,\alpha)). \tag{6.1.38}$$

But, by (6.1.18),

$$\lim_{\alpha\to 0} w(\boldsymbol{H}(\cdot,\alpha)) = \int_{-\infty}^{+\infty} \dot{\boldsymbol{H}}(t)\cdot\mathscr{G}(+\infty)\boldsymbol{H}(t)\,dt \tag{6.1.39}$$

which means, since $\mathscr{G}(+\infty)$ is symmetric and $\boldsymbol{H}(-\infty)=\boldsymbol{E}(+\infty)$ and $\boldsymbol{H}(+\infty)=\boldsymbol{O}$ that

$$\lim_{\alpha\to 0} w(\boldsymbol{H}(\cdot,\alpha)) = -\tfrac{1}{2}\boldsymbol{E}(+\infty)\cdot\mathscr{G}(+\infty)\boldsymbol{E}(+\infty). \tag{6.1.40}$$

The required inequality (6.1.34) now follows on letting $\alpha \to 0$ in (6.1.38) and using (6.1.40), which completes the proof of *II* (i).

Next we turn to proving *II* (ii). We have just proved that if $\mathscr{G}(\cdot)$ is compatible with thermodynamics then $\mathscr{G}(+\infty)$ is symmetric and the inequality (6.1.34) holds provided $\boldsymbol{E}(\cdot)$ starts from the virgin state. However (6.1.34) can be written as

$$\int_{-\infty}^{+\infty} \left\{\mathscr{G}(0)\boldsymbol{E}(t) + \int_0^{+\infty} \dot{\mathscr{G}}(s)\boldsymbol{E}(t-s)\,ds\right\}\cdot\dot{\boldsymbol{E}}(t)\,dt \geqslant \int_{-\infty}^{+\infty} \dot{\boldsymbol{E}}(t)\cdot\mathscr{G}(+\infty)\boldsymbol{E}(t)\,dt,$$

that is as

$$\int_{-\infty}^{+\infty} \left\{(\mathscr{G}(0)-\mathscr{G}(+\infty))\boldsymbol{E}(t) + \int_0^{+\infty} \frac{d}{ds}(\mathscr{G}(s)-\mathscr{G}(+\infty))\boldsymbol{E}(s)\,ds\right\}\cdot\dot{\boldsymbol{E}}(t)\,dt \geqslant 0,$$

which means that the relaxation function $\mathscr{G}(\cdot)-\mathscr{G}(+\infty)$ is dissipative and the 'only if' part of *II* (ii) is proved.

To prove the 'if' part of *II* (ii) let us suppose that $\mathscr{G}(\cdot)-\mathscr{G}(+\infty)$ is dissipative and that $\mathscr{G}(+\infty)$ is symmetric. If $\boldsymbol{E}(\cdot)$ is any closed path and if $\hat{\boldsymbol{E}}(\cdot)$ is the closed path starting from the virgin state associated with it as in (6.1.31) then (6.1.32) tells us that

$$w(\boldsymbol{E}(\cdot)) = w(\hat{\boldsymbol{E}}(\cdot)). \tag{6.1.41}$$

However, since $\hat{\boldsymbol{E}}(\cdot)$ starts from the virgin state the work done on $\hat{\boldsymbol{E}}(\cdot)$, computed using the relaxation function $\mathscr{G}(\cdot)-\mathscr{G}(+\infty)$, cannot be negative, that is to say

$$\int_{-\infty}^{+\infty} \left\{(\mathscr{G}(0)-\mathscr{G}(+\infty))\boldsymbol{E}(t) + \int_0^{+\infty} \frac{d}{ds}(\mathscr{G}(s)-\mathscr{G}(+\infty))\hat{\boldsymbol{E}}(t-s)\,ds\right\}\cdot\hat{\boldsymbol{E}}(t)\,dt \geqslant 0,$$

which implies, on rearranging, that

$$w(\hat{\boldsymbol{E}}(\cdot)) \geqslant \int_{-\infty}^{+\infty} \hat{\boldsymbol{E}}(t) \cdot \mathcal{G}(+\infty) \hat{\boldsymbol{E}}(t) \, dt. \tag{6.1.42}$$

However, since $\hat{\boldsymbol{E}}(\cdot)$ is closed and $\mathcal{G}(+\infty)$ is symmetric the right-hand side of (6.1.42) vanishes and it follows from (6.1.41) that

$$w(\boldsymbol{E}(\cdot)) \geqslant 0,$$

that is $\mathcal{G}(\cdot)$ is compatible with thermodynamics and the proof of *II* is complete.

As we pointed out previously it follows from *I* and *II* that relaxation functions which are compatible with thermodynamics are subject to the restrictions *III*. An example of such a relaxation function will be considered in section 6.2.

We have already shown that a dissipative relaxation function is subject to the inequality (6.1.28). It follows from *II* (ii) that if $\mathcal{G}(\cdot)$ is compatible with thermodynamics then the corresponding inequality obtained by replacing $\mathcal{G}(\cdot)$ by $\mathcal{G}(\cdot) - \mathcal{G}(+\infty)$, namely

$$\tfrac{1}{2}\boldsymbol{A} \cdot (\mathcal{G}(0) - \mathcal{G}(+\infty))\boldsymbol{A} + \boldsymbol{B} \cdot (\mathcal{G}(s) - \mathcal{G}(\infty))\boldsymbol{A} + \tfrac{1}{2}\boldsymbol{B} \cdot (\mathcal{G}(0) - \mathcal{G}(+\infty))\boldsymbol{B} \geqslant 0$$

must hold for every $s > 0$ and for every pair of symmetric tensors $\boldsymbol{A}$, $\boldsymbol{B}$. If we replace $\boldsymbol{A}$ by $\alpha \boldsymbol{A}$ and $\boldsymbol{B}$ by $\beta \boldsymbol{B}$ where α, β are any scalars we can deduce that

$$\begin{aligned} \{\boldsymbol{B} \cdot (\mathcal{G}(s) - \mathcal{G}(+\infty))\boldsymbol{A}\}^2 &\leqslant \{\boldsymbol{A} \cdot (\mathcal{G}(0) - \mathcal{G}(+\infty))\boldsymbol{A}\}\,\{\boldsymbol{B} \cdot (\mathcal{G}(0) - \mathcal{G}(+\infty))\boldsymbol{B}\} \\ &\leqslant \|\mathcal{G}(0) - \mathcal{G}(+\infty)\|^2 \, \|\boldsymbol{A}\|^2 \, \|\boldsymbol{B}\|^2, \end{aligned}$$

and if we now take

$$\boldsymbol{B} = (\mathcal{G}(s) - \mathcal{G}(+\infty))\boldsymbol{A}$$

it is a straightforward matter to deduce that if $\mathcal{G}(\cdot)$ *is compatible with thermodynamics then*

$$\|\mathcal{G}(s) - \mathcal{G}(+\infty)\| \leqslant \|\mathcal{G}(0) - \mathcal{G}(+\infty)\|, \qquad s \geqslant 0. \tag{6.1.43}$$

This inequality has an interesting consequence for linear elastic materials[1]: a *linear elastic material* is a linear viscoelastic material whose relaxation function is identically constant, that is $\mathcal{G}(s) = \mathcal{G}(0)$ for every $s \geqslant 0$. The elastic moduli $\mathcal{G}(0)$ and $\mathcal{G}(+\infty)$ of an elastic material necessarily coincide. Moreover the inequality (6.1.43) tells us that if $\mathcal{G}(0) = \mathcal{G}(+\infty)$ then $\mathcal{G}(s) = \mathcal{G}(+\infty) = \mathcal{G}(0)$ for every $s \geqslant 0$ and so *a linear viscoelastic material which is compatible with thermodynamics is elastic if and only if its equilibrium and instantaneous elastic moduli coincide.*

---

[1] See [39].

In particular, since the response of the material in equilibrium is governed by the modulus $\mathscr{G}(+\infty)$ and the response to rapid changes in strain by the modulus $\mathscr{G}(0)$ it follows that *for a linear viscoelastic material which is compatible with thermodynamics and which has a genuine memory for the past, in the sense that it is not elastic, its response to rapid changes in strain can never be the same as its response in equilibrium.* Incidentally we have recovered the familiar result that *the elastic modulus of a linear elastic material which is compatible with thermodynamics is symmetric.*

6.2 The Symmetry of the Relaxation Function

We have seen that as a consequence of compatibility with thermodynamics, and also of the stronger requirement of dissipativity, the elastic moduli must both be symmetric. It is sometime asserted[1] on the grounds of Onsager's reciprocal relations that in fact $\mathscr{G}(s)$ must be symmetric not only for $s=0$ and $s=+\infty$ but for every s, i.e.

$$\mathscr{G}_{ijkl}(s)=\mathscr{G}_{klij}(s), \qquad 0\leqslant s\leqslant+\infty\,. \tag{6.2.1}$$

It would be very interesting to know if this relation is true for real materials[2]; its truth or falsity however is certainly not a consequence of compatibility with thermodynamics or even of the requirement that $\mathscr{G}(\cdot)$ be dissipative because examples of dissipative relaxation functions can be constructed[3] in such a way that $\mathscr{G}(s)$ is not symmetric for any s in $0<s<+\infty$. Indeed *if $\mathscr{A}$ is a symmetric and positive semi-definite tensor, if the tensor $\mathscr{B}$ is positive definite, if it commutes with its transpose, i.e. $\mathscr{B}\mathscr{B}^{\mathrm{T}}=\mathscr{B}^{\mathrm{T}}\mathscr{B}$ and if the skew tensor $\mathscr{B}-\mathscr{B}^{\mathrm{T}}$ has two non-zero pure imaginary eigenvalues $i\beta_1, i\beta_2$ whose ratio β_1/β_2 is irrational then the relaxation function*

$$\mathscr{G}(s)=\mathscr{A}+\exp(-s\mathscr{B}), \qquad s\geqslant 0 \tag{6.2.2}$$

is dissipative, and hence compatible with thermodynamics, but $\mathscr{G}(s)$ is not symmetric for any s in $0<s<+\infty$.[4]

---

[1] References to claims of this sort can be found in the paper of Rogers and Pipkin [64].

[2] In [64] Rogers and Pipkin have suggested possible experiments for determining wheter $\mathscr{G}(\cdot)$ is symmetric or not.

[3] The first counterexample of this kind is due to Shu and Onat [65].

[4] The requirements on $\mathscr{B}$ can all be satisfied: e.g. we can choose a skew fourth order tensor $\boldsymbol{\Omega}$ in such a way that $\boldsymbol{\Omega}$ is invertible and $\boldsymbol{\Omega}$ has two eigenvalues whose ratio is irrational. If we set

$$\mathscr{B}=c(-\boldsymbol{\Omega}^2)^m+\boldsymbol{\Omega}$$

where $c>0$ and $m\geqslant 0$ is an integer then $\mathscr{B}$ meets our requirements.

To prove that this is so we begin by remarking that when $\mathscr{G}(\cdot)$ is defined as in (6.2.2)

$$\|\mathscr{G}(s)-\mathscr{A}\| = O(\exp(-\beta s))$$

and

$$\|\dot{\mathscr{G}}(s)\| = O(\exp(-\beta s))$$

where $\beta>0$ is the smallest eigenvalue of the positive definite and symmetric tensor $\frac{1}{2}(\mathscr{B}+\mathscr{B}^T)$. *Thus* $\mathscr{G}(+\infty)=\mathscr{A}$, $\mathscr{G}(0)-\mathscr{G}(+\infty)$ is the identity tensor and the integrals

$$\int_0^{+\infty} \|\dot{\mathscr{G}}(s)\|\, ds, \qquad \int_0^{+\infty} s\|\dot{\mathscr{G}}(s)\|\, ds$$

are finite.

Next we prove that $\mathscr{G}(\cdot)$ is dissipative. Suppose that $\boldsymbol{E}(\cdot)$ is any path starting from the virgin state and that $\boldsymbol{E}(t)\equiv 0$ for $t\leqslant t_0$ and $\boldsymbol{E}(t)\equiv \boldsymbol{E}(+\infty)$ for $t\geqslant t_1$. The hereditary law (6.1.1) tells us that the stress on this path is

$$\boldsymbol{T}(t)=\mathscr{A}\,\boldsymbol{E}(t)+\hat{\boldsymbol{T}}(t) \tag{6.2.3}$$

where

$$\hat{\boldsymbol{T}}(t) = \boldsymbol{E}(t) - \int_{-\infty}^{t} \{\exp((s-t)\mathscr{B})\}\,\mathscr{B}\,\boldsymbol{E}(s)\, ds. \tag{6.2.4}$$

On differentiating (6.2.4) throughout with respect to t we find the relation

$$\dot{\boldsymbol{E}}(t) = \dot{\hat{\boldsymbol{T}}}(t)+\mathscr{B}\,\hat{\boldsymbol{T}}(t)$$

and on using this relation, the fact that $\boldsymbol{E}(-\infty)=\boldsymbol{O}$, the fact that $\hat{\boldsymbol{T}}(t_0)=\boldsymbol{O}$ and the properties of $\mathscr{A}$ and $\mathscr{B}$ we find that the work done satisfies

$$\begin{aligned}
w(\boldsymbol{E}(\cdot)) &= \int_{t_0}^{t_1} \boldsymbol{T}(t)\cdot\dot{\boldsymbol{E}}(t)\,dt \\
&= \int_{t_0}^{t_1} \dot{\boldsymbol{E}}(t)\cdot\mathscr{A}\,\boldsymbol{E}(t)\,dt + \int_{t_0}^{t_1} \hat{\boldsymbol{T}}(t)\cdot(\dot{\hat{\boldsymbol{T}}}(t)+\mathscr{B}\,\hat{\boldsymbol{T}}(t))\,dt \\
&= \tfrac{1}{2}\boldsymbol{E}(+\infty)\cdot\mathscr{A}\,\boldsymbol{E}(+\infty)+\tfrac{1}{2}\hat{\boldsymbol{T}}(t_1)\cdot\hat{\boldsymbol{T}}(t_1) + \int_{t_0}^{t_1} \hat{\boldsymbol{T}}(t)\cdot\mathscr{B}\,\hat{\boldsymbol{T}}(t)\,dt \geqslant 0,
\end{aligned}$$

which means that $\mathscr{G}(\cdot)$ is dissipative.

Moreover it is not difficult to check that $\mathscr{G}(s)$ cannot be symmetric for any s in $0<s<+\infty$ for if it were symmetric we should have

$$\exp(-s\mathscr{B}) = \exp(-s\mathscr{B}^T)$$

which would mean, because $\mathscr{B}$ and $\mathscr{B}^T$ commute, that $\exp(-s(\mathscr{B}-\mathscr{B}^T))$ is the identity tensor and hence that the eigenvalues of $-s(\mathscr{B}-\mathscr{B}^T)$

are numbers of the form $2in\pi$ where n is an integer, which may be positive, negative or zero. But this contradicts the assumption made about the skew tensor $\mathscr{B}-\mathscr{B}^T$ because that assumption ensures that $-s(\mathscr{B}-\mathscr{B}^T)$ has two non-zero eigenvalues $-is\beta_1$ and $-is\beta_2$ whose ratio β_1/β_2 is irrational and so $\mathscr{G}(s)$ can never be symmetric for $0<s<+\infty$.

Although the symmetry of $\mathscr{G}(s)$, $0<s<+\infty$, required by the Onsager reciprocal relations is not a consequence of compatibility with thermodynamics it can be characterised[1] in terms of a different condition which involves the work done on certain paths and the idea of the time-reversal of a path. This kind of characterisation is interesting in view of the claim which is commonly made[2] that the Onsager relations reflect 'time-reversal invariance' or 'microscopic reversibility'.

The *time-reversal* of a path $\boldsymbol{E}(\cdot)$ is the path $\overline{\boldsymbol{E}}(\cdot)$ for which $\overline{\boldsymbol{E}}(t)=\boldsymbol{E}(-t)$, $-\infty<t<+\infty$. We shall prove *$\mathscr{G}(t)$ is symmetric for every t in $0\leqslant t\leqslant+\infty$ if and only if the work done on every closed path $\boldsymbol{E}(\cdot)$ starting from the virgin state is invariant under time-reversal, i.e.*

$$w(\boldsymbol{E}(\cdot))=w(\overline{\boldsymbol{E}}(\cdot))\,. \tag{6.2.5}$$

It is a straightforward matter to check the necessity of (6.2.5). For if $\boldsymbol{E}(\cdot)$ is closed and starts from the virgin state

$$w(\boldsymbol{E}(\cdot))=\int_{-\infty}^{+\infty}\int_{-\infty}^{t}\dot{\boldsymbol{E}}(t)\cdot\mathscr{G}(t-s)\dot{\boldsymbol{E}}(s)\,ds\,dt\,. \tag{6.2.6}$$

Its time-reversal $\overline{\boldsymbol{E}}(\cdot)$ is also closed and starts from the virgin state and

$$\begin{aligned} w(\overline{\boldsymbol{E}}(\cdot)) &= \int_{-\infty}^{+\infty}\int_{-\infty}^{t}\dot{\overline{\boldsymbol{E}}}(t)\cdot\mathscr{G}(t-s)\dot{\overline{\boldsymbol{E}}}(s)\,ds\,dt \\ &= \int_{-\infty}^{+\infty}\int_{-\infty}^{t}\dot{\boldsymbol{E}}(-t)\cdot\mathscr{G}(t-s)\dot{\boldsymbol{E}}(-s)\,ds\,dt\,. \end{aligned}$$

On making the change of variables $s'=-s$, $t'=-t$ we find that

$$w(\overline{\boldsymbol{E}}(\cdot))=\int_{-\infty}^{+\infty}\int_{t'}^{+\infty}\dot{\boldsymbol{E}}(t')\cdot\mathscr{G}(s'-t')\dot{\boldsymbol{E}}(s')\,ds'\,dt'$$

and if we interchange the orders of integration, as we may, we deduce that

$$\begin{aligned} w(\overline{\boldsymbol{E}}(\cdot)) &= \int_{-\infty}^{+\infty}\int_{-\infty}^{s'}\dot{\boldsymbol{E}}(t')\cdot\mathscr{G}(s'-t')\dot{\boldsymbol{E}}(s')\,dt'\,ds' \\ &= \int_{-\infty}^{+\infty}\int_{-\infty}^{s'}\dot{\boldsymbol{E}}(s')\cdot\mathscr{G}^T(s'-t')\dot{\boldsymbol{E}}(t')\,dt'\,ds'\,, \end{aligned}$$

---

[1] See Day [37].

[2] See, for example, Ch. IV, § 3 of the treatise of de Groot and Mazur [45].

that is to say

$$w(\bar{\boldsymbol{E}}(\cdot))=\int_{-\infty}^{+\infty}\int_{-\infty}^{t}\dot{\boldsymbol{E}}(t)\cdot\mathscr{G}^T(t-s)\dot{\boldsymbol{E}}(s)\,d s\,d t\,. \tag{6.2.7}$$

A comparison of (6.2.6) and (6.2.7) shows that (6.2.5) holds if $\mathscr{G}(t)$ is symmetric for every t in $0\leqslant t<+\infty$.

Conversely, suppose that (6.2.5) does hold. Then the function

$$\mathscr{L}(\cdot)=\mathscr{G}(\cdot)-\mathscr{G}^T(\cdot) \tag{6.2.8}$$

whose values are skew tensors satisfies

$$\int_{-\infty}^{+\infty}\int_{-\infty}^{t}\dot{\boldsymbol{E}}(t)\cdot\mathscr{L}(t-s)\dot{\boldsymbol{E}}(s)\,d s\,d t=0 \tag{6.2.9}$$

for every closed path $\boldsymbol{E}(\cdot)$ starting from the virgin state. If we can show that $\mathscr{L}(t)=\boldsymbol{O}$ for every t in $0\leqslant t\leqslant+\infty$ the symmetry of $\mathscr{G}(t)$ follows.

Let $t_1>0, t_2>0$ be arbitrary but fixed. Then for any ε in $0<\varepsilon<\min(t_1,t_2)$ and for any symmetric tensors $\boldsymbol{A}$, $\boldsymbol{B}$ we can define a piecewise linear closed path $\boldsymbol{E}(\cdot)$ starting from the virgin state by setting $\boldsymbol{E}(t)=\boldsymbol{O}$ for $t\leqslant 0$, $\boldsymbol{E}(t)=(1/\varepsilon)t\boldsymbol{A}$ for $0\leqslant t\leqslant\varepsilon$, $\boldsymbol{E}(t)=\boldsymbol{A}$ for $\varepsilon\leqslant t\leqslant t_1$, $\boldsymbol{E}(t)=\boldsymbol{A}+(1/\varepsilon)(t-t_1)(\boldsymbol{B}-\boldsymbol{A})$ for $t_1\leqslant t\leqslant t_1+\varepsilon$, $\boldsymbol{E}(t)=\boldsymbol{B}$ for $t_1+\varepsilon\leqslant t\leqslant t_1+t_2$, $\boldsymbol{E}(t)=(1/\varepsilon)\,(t_1+t_2+\varepsilon-t)\boldsymbol{B}$ for $t_1+t_2\leqslant t\leqslant t_1+t_2+\varepsilon$, and $\boldsymbol{E}(t)=\boldsymbol{O}$ for $t\geqslant t_1+t_2+\varepsilon$ (see Fig. 8). For this path (6.2.9) tells us that

$$\begin{aligned}
0=&\frac{1}{\varepsilon^2}\int_0^{\varepsilon}\int_0^{t}\boldsymbol{A}\cdot\mathscr{L}(t-s)\boldsymbol{A}\,d s\,d t+\frac{1}{\varepsilon^2}\int_{t_1}^{t_1+\varepsilon}\int_0^{\varepsilon}(\boldsymbol{B}-\boldsymbol{A})\cdot\mathscr{L}(t-s)\boldsymbol{A}\,d s\,d t\\
&+\frac{1}{\varepsilon^2}\int_{t_1}^{t_1+\varepsilon}\int_{t_1}^{t}(\boldsymbol{B}-\boldsymbol{A})\cdot\mathscr{L}(t-s)(\boldsymbol{B}-\boldsymbol{A})\,d s\,d t\\
&-\frac{1}{\varepsilon^2}\int_{t_1+t_2}^{t_1+t_2+\varepsilon}\int_0^{\varepsilon}\boldsymbol{B}\cdot\mathscr{L}(t-s)\boldsymbol{A}\,d s\,d t\\
&-\frac{1}{\varepsilon^2}\int_{t_1+t_2}^{t_1+t_2+\varepsilon}\int_{t_1}^{t_1+\varepsilon}\boldsymbol{B}\cdot\mathscr{L}(t-s)(\boldsymbol{B}-\boldsymbol{A})\,d s\,d t\\
&+\frac{1}{\varepsilon^2}\int_{t_1+t_2}^{t_1+t_2+\varepsilon}\int_{t_1+t_2}^{t}\boldsymbol{B}\cdot\mathscr{L}(t-s)\boldsymbol{B}\,d s\,d t\,,
\end{aligned}$$

which simplifies, because $\mathscr{L}(\cdot)$ is skew, to

$$0=\frac{1}{\varepsilon^2}\int_{t_1}^{t_1+\varepsilon}\int_0^{\varepsilon}\boldsymbol{B}\cdot\mathscr{L}(t-s)\boldsymbol{A}\,ds\,dt-\frac{1}{\varepsilon^2}\int_{t_1+t_2}^{t_1+t_2+\varepsilon}\int_0^{\varepsilon}\boldsymbol{B}\cdot\mathscr{L}(t-s)\boldsymbol{A}\,ds\,dt$$

$$+\frac{1}{\varepsilon^2}\int_{t_1+t_2}^{t_1+t_2+\varepsilon}\int_{t_1}^{t_1+\varepsilon}\boldsymbol{B}\cdot\mathscr{L}(t-s)\boldsymbol{A}\,ds\,dt\,.$$

If we now let $\varepsilon\to 0$ we deduce that

$$0=\boldsymbol{B}\cdot(\mathscr{L}(t_1)-\mathscr{L}(t_1+t_2)+\mathscr{L}(t_2))\boldsymbol{A}$$

for all symmetric tensors $\boldsymbol{A}$, $\boldsymbol{B}$, and hence

$$\mathscr{L}(t_1+t_2)=\mathscr{L}(t_1)+\mathscr{L}(t_2) \tag{6.2.10}$$

for every $t_1>0$ and for every $t_2>0$. The only continuous solutions of (6.2.10) satisfy

$$\mathscr{L}(t)=t\,\mathscr{L}(1), \qquad t>0\,. \tag{6.2.11}$$

However, our assumptions on the continuous function $\mathscr{G}(\cdot)$ ensure that the limits $\lim_{s\to+\infty}\mathscr{G}(s)$ and $\lim_{s\to+\infty}\mathscr{G}^T(s)$ exist. Consequently the functions $\|\mathscr{G}(\cdot)\|$ and $\|\mathscr{G}^T(\cdot)\|$ are bounded and so is $\|\mathscr{L}(\cdot)\|$, which contradicts

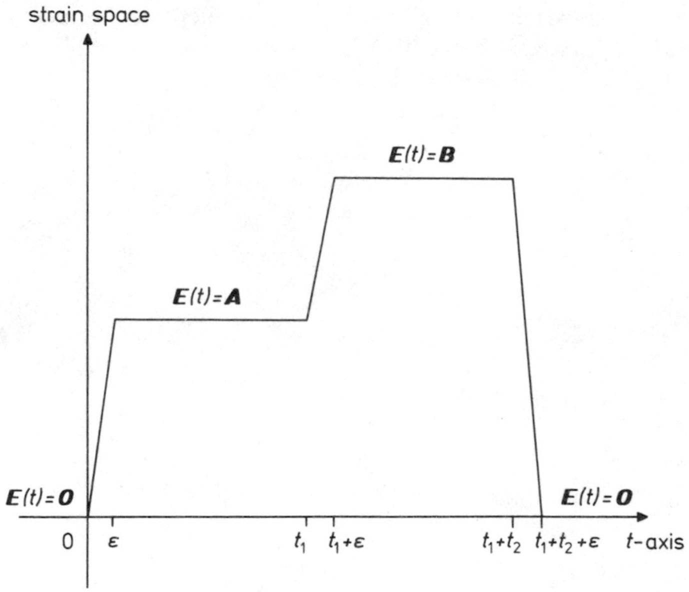

Fig. 8

(6.2.11) unless $\mathscr{L}(1)=\boldsymbol{O}$. Hence $\mathscr{L}(t)=\boldsymbol{O}$ for every t in $0<t<+\infty$ and the symmetry of $\mathscr{G}(\cdot)$ follows.

6.3 A Remark on the Monotonicity of Relaxation Functions

Experimentally the one-dimensional relaxation functions of linear viscoelastic materials are found to be monotone decreasing functions of the time. It is natural to ask if this property is a consequence of compatibility with thermodynamics or of the stronger dissipative requirement. The answer to this question is 'no' for Gurtin and Herrera [49] have constructed a dissipative one-dimensional relaxation function which is not monotone decreasing. It turns out that there is a criterion, formulated once again in terms of the work done on certain strain paths, which does force analytic one-dimensional relaxation functions to be strictly monotone (see [35]); this criterion is of course logically distinct from compatibility with thermodynamics, from dissipativity and from the condition which guarantees symmetry, namely invariance of the work done under time-reversal.

If one takes the view that the restrictions found in 6.1, the monotonicity of one-dimensional relaxation functions and the symmetry of the relaxation functions of anisotropic linear viscoelastic materials, if experiments show them to be symmetric, should all be consequences of a second law of thermodynamics then it is clear that neither the thermodynamic inequality of Chapter 2 nor the Clausius-Duhem inequality can be regarded as the definitive form of the second law.

References

1. Breuer, S.: Lower bounds on work in linear viscoelasticity. Quart. Appl. Math. **27,** 139–146 (1969).
2. Breuer, S., Onat, E. T.: On recoverable work in linear viscoelasticity. Z. Angew. Math. Phys. **15,** 12–21 (1964).
3. Breuer, S., Onat, E. T.: On the determination of free energy in linear viscoelastic solids. Z. Angew. Math. Phys. **15,** 184–191 (1964).
4. Chen, P. J., Gurtin, M. E.: On a theory of heat conduction involving two temperatures. Z. Angew. Math. Phys. **19,** 614–627 (1968).
5. Chen, P. J., Gurtin, M. E., Williams, W. O.: A note on non-simple heat conduction. Z. Angew. Math. Phys. **19,** 969–970 (1968).
6. Chen, P. J., Gurtin, M. E., Williams, W. O.: On the thermodynamics of non-simple elastic materials with two temperatures. Z. Angew. Math. Phys. **20,** 107–112 (1969).
7. Coleman, B. D.: Kinematical concepts with applications in the mechanics and thermodynamics of incompressible viscoelastic fluids. Arch. Rational Mech. Anal. **9,** 273–300 (1962).
8. Coleman, B. D.: Thermodynamics of materials with memory. Arch. Rational Mech. Anal. **17,** 1–46 (1964).
9. Coleman, B. D.: On thermodynamics, strain impulses and viscoelasticity. Arch. Rational Mech. Anal. **17,** 230–254 (1964).
10. Coleman, B. D.: On the stability of equilibrium states of general fluids. Arch. Rational Mech. Anal. **36,** 1–32 (1970).
11. Coleman, B. D., Dill, E. H.: On the stability of certain motions of incompressible materials with memory. Arch. Rational Mech. Anal. **30,** 197–224 (1970).
12. Coleman, B. D., Greenberg, J. M.: Thermodynamics and the stability of fluid motion. Arch. Rational Mech. Anal. **25,** 321–341 (1967).
13. Coleman, B. D., Greenberg, J. M., Gurtin, M. E.: Waves in materials with memory V. On the amplitude of acceleration waves and mild discontinuities. Arch. Rational Mech. Anal. **22,** 333–354 (1966).
14. Coleman, B. D., Gurtin, M. E.: Thermodynamics and wave propagation. Quart. Appl. Math. **24,** 257–262 (1966).
15. Coleman, B. D., Gurtin, M. E.: Thermodynamics with internal state variables. J. Chem. Phys. **47,** 597–613 (1967).

16. Coleman, B. D., Gurtin, M. E.: Equipresence and constitutive equations for rigid heat conductors. Z. Angew. Math. Phys. **18,** 199–208 (1967).

17. Coleman, B. D., Gurtin, M. E., Herrera, R. I.: Wave Propagation in Dissipative Materials. Berlin–Heidelberg–New York: Springer 1965.

18. Coleman, B. D., Mizel, V. J.: Thermodynamics and departures from Fourier's law of heat conduction. Arch. Rational Mech. Anal. **13,** 245–261 (1963).

19. Coleman, B. D., Mizel, V. J.: Existence of caloric equations of state in thermodynamics. J. Chem. Phys. **40,** 1116–1125 (1964).

20. Coleman, B. D., Mizel, V. J.: Norms and semi-groups in the theory of fading memory. Arch. Rational Mech. Anal. **23,** 87–123 (1966).

21. Coleman, B. D., Mizel, V. J.: A general theory of dissipation in materials with memory. Arch. Rational Mech. Anal. **27,** 255–274 (1968).

22. Coleman, B. D., Mizel, V. J.: On the general theory of fading memory. Arch. Rational Mech. Anal. **29,** 18–31 (1968).

23. Coleman, B. D., Mizel, V. J.: On thermodynamic conditions for the stability of evolving systems. Arch. Rational Mech. Anal. **29,** 105–113 (1968).

24. Coleman, B. D., Mizel, V. J.: On the stability of solutions of functional-differential equations. Arch. Rational Mech. Anal. **30,** 173–196 (1968).

25. Coleman, B. D., Noll, W.: An approximation theorem for functionals with applications in continuum mechanics. Arch. Rational Mech. Anal. **6,** 355–370 (1960).

26. Coleman, B. D., Noll, W.: Foundations of linear viscoelasticity. Rev. Modern Phys. **33,** 239–249 (1961).

27. Coleman, B. D., Noll, W.: The thermodynamics of elastic materials with heat conduction and viscosity. Arch. Rational Mech. Anal. **13,** 167–178 (1963).

28. Coleman, B. D., Owen, D. R.: On the thermodynamics of materials with memory. Arch. Rational Mech. Anal. **36,** 245–269 (1970).

29. Day, W. A.: Thermodynamics based on a work axiom. Arch. Rational Mech. Anal. **31,** 1–34 (1968).

30. Day, W. A.: A theory of thermodynamics for materials with memory. Arch. Rational Mech. Anal. **34,** 85–96 (1969).

31. Day, W. A.: Useful strain histories in linear viscoelasticity. Quart. Appl. Math. **27,** 255–259 (1969).

32. Day, W. A.: A note on useful work. Quart. Appl. Math. **27,** 260–262 (1969).

33. Day, W. A.: On the conversion of heat into mechanical work. J. Math. Anal. Appl. **27,** 210–224 (1969).

34. Day, W. A.: Reversibility, recoverable work and free energy in linear viscoelasticity. Quart. J. Mech. Appl. Math. **23,** 1–15 (1970).

35. Day, W. A.: On monotonicity of the relaxation functions of viscoelastic materials. Proc. Cambridge Philos. Soc. **67,** 503–508 (1970).

36. Day, W. A.: Some results on the least work needed to produce a given strain in a given time in a viscoelastic material and a uniqueness theorem for dynamic viscoelasticity. Quart. J. Mech. Appl. Math. **23,** 469–479 (1970).

37. Day, W. A.: Time-reversal and the symmetry of the relaxation function of a linear viscoelastic material. Arch. Rational Mech. Anal. **40,** 155–159 (1971).

38. Day, W.A.: Restrictions on relaxation functions in linear viscoelasticity. Quart. J. Mech. Appl. Math. **24**, 487–497 (1971).

39. Day, W.A.: When is a linear viscoelastic material elastic? Mathematika. **18,** 134–137 (1971).

40. Day, W. A., Gurtin, M. E.: On the symmetry of the conductivity tensor and other restrictions in the non-linear theory of heat conduction. Arch. Rational Mech. Anal. **33,** 26-32 (1969).

41. Fisher, G. M. C., Leitmann, M.: On continuum thermodynamics with surfaces. Arch. Rational Mech. Anal. **30,** 225–262 (1968).

42. Green, A. E., Laws, N.: On the formulation of constitutive equations in thermodynamical theories of continua. Quart. J. Mech. Appl. Math. **20,** 265–275 (1967).

43. Green, A. E., Laws, N.: On a global entropy production inequality. To be published.

44. Green, A. E., Naghdi, P. M.: A general theory of an elastic-plastic continuum. Arch. Rational Mech. Anal. **18,** 251–281 (1965).

45. de Groot, S. R., Mazur, P.: Non-equilibrium Thermodynamics. Amsterdam, North-Holland 1962.

46. Gurtin, M. E.: Thermodynamics and the possibility of spatial interaction in rigid heat conductors. Arch. Rational Mech. Anal. **18,** 335–342 (1965).

47. Gurtin, M. E.: Thermodynamics and the possibility of spatial interaction in elastic materials. Arch. Rational Mech. Anal. **19,** 339–352 (1965).

48. Gurtin, M. E.: On the thermodynamics of materials with memory. Arch. Rational Mech. Anal. **28,** 40–50 (1968).

49. Gurtin, M. E., Herrera, I.: On dissipation inequalities and linear viscoelasticity. Quart. Appl. Math. **23,** 235–245 (1965).

50. Gurtin, M. E., Sternberg, E.: On the linear theory of viscoelasticity. Arch. Rational Mech. Anal. **11,** 291–356 (1962).

51. Gurtin, M. E., Williams, W. O.: On the inclusion of the complete symmetry group in the unimodular group. Arch. Rational Mech. Anal. **23,** 163–172 (1966).

52. Gurtin, M. E., Williams, W. O.: On the Clausius-Duhem inequality. Z. Angew. Math. Phys. **17,** 626–633 (1966).

53. Gurtin, M. E., Williams, W. O.: An axiomatic foundation for continuum thermodynamics. Arch. Rational Mech. Anal. **26,** 83–117 (1967).

54. Koh, S. L., Eringen, A. C.: On the foundations of non-linear thermoviscoelasticity. Internat. J. Engrg. Sci. **1,** 199–229 (1963).

55. König, H., Meixner, J.: Lineare Systeme und lineare Transformationen. Math. Nachr. **19,** 256–322 (1958).

56. Laws, N.: On the thermodynamics of certain materials with memory. Internat. J. Engrg. Sci. **5,** 427–434 (1967).

57. Martin, J. B., Ponter, A. R. S.: A note on a work inequality in linear viscoelasticity. Quart. Appl. Math. **24,** 161–165 (1966).

58. Meixner, J.: Processes in simple thermodynamic materials. Arch. Rational Mech. Anal. **33,** 33–53 (1969).

59. Mizel, V. J., Wang, C.-C.: A fading memory hypothesis which suffices for chain-rules. Arch. Rational Mech. Anal. **23,** 124–134 (1966).

60. Müller, I.: On the entropy inequality. Arch. Rational Mech. Anal. **26,** 118–141 (1967).

61. Owen, D. R.: Thermodynamics of materials with elastic range. Arch. Rational Mech. Anal. **31,** 91–112 (1968).

62. Owen, D. R.: A mechanical theory of materials with elastic range. Arch. Rational Mech. Anal. **37,** 85–110 (1970).

63. Pipkin, A. C., Rivlin, R. S.: The formulation of constitutive equations in continuum physics. Technical report DA 4531/4 to the Department of the Army, Ordnance Corps 1958.

64. Rogers, T. G., Pipkin, A. C.: Asymmetric relaxation and compliance matrices in linear viscoelasticity. Z. Angew. Math. Phys. **14,** 334–343 (1963).

65. Shu, L. S., Onat, E. T.: On anisotropic linear viscoelastic solids. Proceedings of the Fourth Symposium on Naval Structural Mechanics, Purdue University, April 1965. Reprinted in Mechanics and Chemistry of Solid Propellants. Oxford and New York: Pergamon 1966.

66. Truesdell, C.: The mechanical foundations of elasticity and fluid dynamics. J. Rational Mech. Anal. **1,** 125–300 (1952).

67. Truesdell, C.: Rational Thermodynamics. New York: McGraw-Hill 1969.

68. Truesdell, C., Noll, W.: The Non-linear Field Theories of Mechanics. Handbuch der Physik Vol. III/3 (ed. S. Flügge). Berlin–Heidelberg–New York: Springer 1965.

69. Truesdell, C., Toupin, R. A.: The Classical Field Theories. Handbuch der Physik Vol. III/1 (ed. S. Flügge). Berlin–Göttingen-Heidelberg: Springer 1960.

70. Wang, C.-C.: Stress relaxation and the principle of fading memory. Arch. Rational Mech. Anal. **18,** 117–126 (1965).

71. Wang, C.-C.: The principle of fading memory. Arch. Rational Mech. Anal. **18,** 343–366 (1965).

72. Wang, C.-C., Bowen, R. M.: On the thermodynamics of non-linear materials with quasi-elastic response. Arch. Rational Mech. Anal. **22,** 79–99 (1966).

73. Williams, W. O.: On internal interactions and the concept of thermal isolation. Arch. Rational Mech. Anal. **34,** 245–258 (1969).

74. Williams, W. O.: Thermodynamics of rigid continua. Arch. Rational Mech. Anal. **36,** 270–284 (1970).

Subject Index

Springer Tracts in Natural Philosophy

Vol. 1 Gundersen: Linearized Analysis of One-Dimensional Magnetohydrodynamic Flows. With 10 figures. X, 119 pages. 1964. Cloth DM 22,—

Vol. 2 Walter: Differential- und Integral-Ungleichungen und ihre Anwendung bei Abschätzungs- und Eindeutigkeitsproblemen
Mit 18 Abbildungen. XIV, 269 Seiten. 1964. Gebunden DM 59,—

Vol. 3 Gaier: Konstruktive Methoden der konformen Abbildung
Mit 20 Abbildungen und 28 Tabellen. XIV, 294 Seiten. 1964.
Gebunden DM 68,—

Vol. 4 Meinardus: Approximation von Funktionen und ihre numerische Behandlung
Mit 21 Abbildungen. VIII, 180 Seiten. 1964. Gebunden DM 49,—

Vol. 5 Coleman, Markovitz, Noll: Viscometric Flows of Non-Newtonian Fluids. Theory and Experiment. With 37 figures. XII, 130 pages. 1966. Cloth DM 22,—

Vol. 6 Eckhaus: Studies in Non-Linear Stability Theory
With 12 figures. VIII, 117 pages. 1965. Cloth DM 22,—

Vol. 7 Leimanis: The General Problem of the Motion of Coupled Rigid Bodies About a Fixed Point. With 66 figures. XVI, 337 pages. 1965. Cloth DM 48,—

Vol. 8 Roseau: Vibrations non linéaires et théorie de la stabilité
Avec 7 figures. XII, 254 pages. 1966. Relié DM 39.—

Vol. 9 Brown: Magnetoelastic Interactions
With 13 figures. VIII, 155 pages. 1966. Cloth DM 38,—

Vol. 10 Bunge: Foundations of Physics
With 5 figures. XII, 312 pages. 1967. Cloth DM 56,—

Vol. 11 Lavrentiev: Some Improperly Posed Problems of Mathematical Physics
With 1 figure. VIII, 72 pages. 1967. Cloth DM 20,—

Vol. 12 Kronmüller: Nachwirkung in Ferromagnetika
Mit 92 Abbildungen. XIV, 329 Seiten. 1968. Gebunden DM 62,—

Vol. 13 Meinardus: Approximation of Functions: Theory and Numerical Methods
With 21 figures. VIII, 198 pages. 1967. Cloth DM 54,—

Vol. 14 Bell: The Physics of Large Deformation of Crystalline Solids
With 166 figures. X, 253 pages. 1968. Cloth DM 48,—

Vol. 15 Buchholz: The Confluent Hypergeometric Function with Special Emphasis on its Applications. XVIII, 238 pages. 1969. Cloth DM 64,—

Vol. 16 Slepian: Mathematical Foundations of Network Analysis
XI, 195 pages. 1968. Cloth DM 44,—

Vol. 17 Gavalas: Nonlinear Differential Equations of Chemically Reacting Systems
With 10 figures. IX, 107 pages. 1968. Cloth DM 34,—

Vol. 18 Marti: Introduction to the Theory of Bases
XII, 149 pages. 1969. Cloth DM 32,—

Vol. 19 Knops, Payne: Uniqueness Theorems in Linear Elasticity
IX, 130 pages. 1971. Cloth DM 36,—

Vol. 20 Edelen, Wilson: Relativity and the Question of Discretization in Astronomy
With 34 figures. XII, 186 pages. 1970. Cloth DM 38,—

Vol. 21 McBride: Obtaining Generating Functions
XIII, 100 pages. 1971. Cloth DM 44,—

Vol. 22 Day: The Thermodynamics of Simple Materials with Fading Memory
With 8 figures. X, 134 pages. 1972. Cloth DM 44,—